PENGUIN BOOKS

REAL LIVES, HALF LIVES

Jeremy Hall makes documentary films for Channel 4 and the BBC. He also writes articles for various newspapers and magazines, including the *Observer* and the *New Scientist*. He lives in London with his wife and son.

Real Lives, Half Lives

TALES FROM THE ATOMIC WASTELAND

Jeremy Hall

PENGUIN BOOKS

To Sumaya

PENGUIN BOOKS

Published by the Penguin Group
Penguin Books Ltd, 27 Wrights Lane, London W8 5TZ, England
Penguin Books USA Inc., 375 Hudson Street, New York, New York 10014, USA
Penguin Books Australia Ltd, Ringwood, Victoria, Australia
Penguin Books Canada Ltd, 10 Alcorn Avenue, Toronto, Ontario, Canada M4V 3B2
Penguin Books (NZ) Ltd, 182–190 Wairau Road, Auckland 10, New Zealand

Penguin Books Ltd, Registered Offices: Harmondsworth, Middlesex, England

First published in Penguin Books 1996
1 3 5 7 9 10 8 6 4 2

Typeset in 10/13 Monophoto Baskerville by
Rowland Phototypesetting Ltd, Bury St Edmunds, Suffolk
Printed in England by Clays Ltd, St Ives plc

Contents

List of Illustrations

Introduction

Flying over the Nevada desert you see a wasteland dotted with mesquite, yucca and Joshua trees. The empty reaches of Frenchman Flats or Yucca Flats, as the adjoining territory is called, go on uninterrupted for miles and miles. It was here, in 1951, that the United States government ordered the first atomic test to be carried out on the American mainland – and, forty years on, the evidence of nuclear destructiveness is plain to see. The desert sand looks baked or vitrified, its surface pock-marked by a hundred craters. The idea for a book about the secret lives of people exposed in one way or another to radioactivity – the victims, the whistleblowers, the protesters, the speculators, the gangsters and terrorists – came to me on a visit to Frenchman Flats in the summer of 1994 as I was standing on the edge of the Sedan crater and gazing into the void. At close quarters this hollow monument to patriotic victory in the Cold War is unquestionably an impressive sight, and yet it also illustrates perhaps too clearly how we mark or, at any rate, despoil our territory at the expense of future generations.

The Sedan crater was the brainchild of nuclear scientist Edward Teller, inventor of the hydrogen bomb and, much later, the chief sponsor of Ronald Reagan's infamous 'Star Wars' policy. Unveiled in 1958 as part of the Plowshare Project, an initiative by the Atomic Energy Commission to find peaceful uses for atomic explosions, Teller came up with a plan to excavate another Panama Canal, or sea-level waterway, across Central America – the Panatomic Canal, as it was dubbed – by detonating a series of 'clean' megaton atomic explosions in order to shift the necessary amounts of earth. Unfortunately – or fortunately, depending on your point of view – the idea was abandoned immediately after the Sedan experiment because it was discovered that the high levels of radioactivity made it impossible to carry out any work inside the crater.

In many ways Teller is an archetypal figure in the history of the nuclear age. A towering genius or a mad boffin, his multifarious projects over five decades, good or bad, yet always incomprehensible to the layperson, reflect a widespread ambivalence towards the power of science and its role in our lives. Indeed it is possible to claim that such ambivalence is now a century old. After all, the nuclear age began a hundred years ago in the Paris laboratory of a physicist by the name of Henri Becquerel.

In 1896, following Wilhelm Roentgen's discovery of the X-ray, Becquerel witnessed a piece of uranium appearing to radiate and to fog a photographic plate. Four years later Ernest Rutherford reported the discovery of a gas emanating from the radioactive element thorium. Marie and Pierre Curie soon discovered that radium (which they had purified from uranium ores in 1898) also gave off a radioactive gas. With the help of Frederick Soddy, an English chemist, Rutherford went on to observe the spontaneous disintegration of radioactive elements, a major discovery of twentieth-century physics. They set about tracing the way uranium, radium and thorium changed their elemental nature by radiating away part of their substance as alpha and beta particles. In the process they discovered that each different radioactive product possessed a characteristic 'half-life', the time needed for its radioactivity to reduce to half its previously measured intensity. The half-life measured the transmutation of half the atoms in the element into atoms of another element or that of a physically variant form of the same element – an 'isotope', as Soddy later named it. Half-life became a way to detect the presence of amounts of transmuted substances – 'decay products' – too small to detect chemically. The half-life of uranium-228 proved to be 4.5 billion years, of radium-226 1,590 years, of one decay product of thorium 22 minutes. Some decay products, however, appeared and transmuted themselves in fractions of a second.

In many ways the story of the first nuclear century is a story of human achievement, of countless benefits to medicine and science. For a hundred years, brilliant minds, from Albert Einstein to Andrei Sakharov, have attempted, often successfully, to exploit the power of the atom. In 1901, for example, Becquerel borrowed a phial from the

Curies containing radium. After carrying it in his waistcoat for six hours, the radium had burned into his skin through several layers of clothing. As a result of this unexpected occurrence, it was deduced that a minute quantity of the element, if placed at the heart of a cancer, would be of sufficient intensity to kill the affected tissue. In later decades, however, the miraculous advances of nuclear medicine were somewhat eclipsed by the glamour of the Manhattan Project and by the development of nuclear weapons.

Nuclear power, too, began as a dream of clean fuel, of 'energy too cheap to meter'. As a by-product, however, the nuclear industry has created millions of tonnes of radioactive material. In writing this book I have tried to explore a fundamental predicament of the nuclear age: the unimaginable difficulty of containing radioactive materials with half-lives that exceed our written history by thousands of years – plutonium, for instance, with a half-life of about 25,000 years.

The dark side of nuclear history is well documented by now. It begins in 1904 with Clarence Dally, Thomas Edison's assistant, reputedly the first person to die as a result of being exposed to radioactivity. The death-toll increased rapidly in the next few years. In 1908 Dr Charles Allen Porter told the American Roentgen Ray Society that there were already fifty cases of radiation sickness. A year later, the president of the society, Dr George Johnson, announced that, 'The insurance companies are beginning to look upon us as undesirable risks.' The most famous death was that of Eben Byers, a good-looking American socialite and amateur golfer, who died in 1932 after drinking a dozen bottles a day of radium water to overcome pain from an arm injury. In the same decade at least fifteen radium dial painters are known to have died as a result of ingesting the radioactive element when they licked their brushes. Two factories in New Jersey and Connecticut closed down in the wake of the scandal that erupted as soon as it was established that radium had given the painters necrosis of the jaw, an incurable disease where the bone marrow dies and the bone literally disintegrates.

Those same desiccated bones were shown to me in the autumn of 1994 by John Russell, curator of the Human Radiobiology Tissue Depository in Spokane, Washington State. His bizarre museum, a nuclear hall of fame or perhaps a chamber of horrors, was set up a quarter of a century ago for the purpose of medical research. Located on the outskirts of the city, this incongruous single-storey building houses the mortal remains of radioactivity's anonymous victims. In addition to the miscellaneous femurs and tibiae of the radium dial painters, the left index finger of Marie Curie is also preserved within the vaults at Spokane. That afternoon, as I wandered up and down the corridors between the humming freezers, I could not help feeling that the dead-tissue depository manifested the human and environmental cost of our hundred-year relationship with the atom. To quote Alvin Weinberg, director of the Oakridge Institute in Tennessee, where the Hiroshima bomb was made, it is clear that we have entered into a Faustian bargain and have no way of going back. There is no escape from the past. The past lives on (and on) in myriad forms of radioactive waste – the ultimate litter – rubbish that will stay rubbish for a very long time.

Radioactivity, of course, presents a unique problem for the senses; it defies our body's warning systems and can only be detected on the dials and liquid-crystal displays of Geiger counters. Nuclear waste takes the form of anything that has become contaminated, from rubber gloves at the local hospital to spent fuel rods discharged from the aptly named 'back end' of the nuclear power process. And yet nowadays, as huge tracts of the natural world are increasingly being restricted and marked 'off limits' for hundreds, even thousands of years, the proliferation of atomic wastelands serves only to highlight that there is another side to the human mastery of nuclear physics, to our hubristic interfering with the fundamental order of nature – and perhaps we have only woken up to this fact as a result of Chernobyl.

Until 26 April 1986 the nuclear station located 100 km north of Kiev, and less than 10 km from Ukraine's border with Belarus, was on the verge of becoming the world's most powerful nuclear dynamo, boasting half a dozen gigawatt reactors capable of providing electricity to 7 million consumers. But in the early hours of the morning on that day, the plant's Block Four reactor exploded, venting at least a hundred times more radiation into the atmosphere than the atomic bomb dropped on Hiroshima. In the wake of the disaster, the Soviet authorities at first claimed that the explosion had led only to a partial meltdown of the reactor. They suggested that military helicopters had been able to douse its blazing core by smothering Block Four with 5,000 tonnes of sand, lead, boron and clay. But scientists have subsequently revealed that the helicopters completely missed their target and that the melted core burned through its protective layers into the lowest levels of the plant basement. If it had penetrated any further into the concrete foundation, it would have come into contact with groundwater and perhaps set off an enormous steam explosion. At the same time it has been proved that Soviet efforts to entomb the reactor in a sarcophagus of concrete and to prevent it from contaminating a nearby river by digging a three-kilometre-long dike, were equally unsuccessful. As it turned out, the explosion at the Chernobyl reactor did in fact lead to a complete core meltdown and to a far worse contamination than was originally reported. The amount of radioactivity unleashed by the world's most notorious nuclear disaster was shown to be at least four or five times greater than the estimates published by the Soviet Union. Flames are reported to have risen thirty metres into the dark sky. In addition, massive amounts of deadly radionuclides were deposited by the wind and rain clouds over 129,500 sq. km of prime farmland in Ukraine, Belarus and Russia. Later on, as the heavily contaminated plume rose in the sky, the radioactivity was carried further afield, across Lithuania and Latvia, into Poland, Sweden and Norway until, finally, it reached parts of Western Europe. The fallout consisted mostly of radioactive caesium-137 and strontium-90, by-products of uranium fission, with thirty-year half-lives that will significantly irradiate the region's soil and food-chain until at least the year 2135.

Ten years on, the governments of Ukraine and the West have only just come to agree upon the terms that will lead to the Chernobyl plant being shut down by the end of the decade. The intervening years have been spent wrangling over money. Then suddenly, at the end of 1995, as a result of the talks in Kiev with officials from the Group of Seven industrialized countries, the Ukrainians volunteered to suspend their demand for US$4 billion as a condition for closing the plant's remaining reactors. Reading the news of this diplomatic breakthrough on a grey morning at home in London, I recalled the trip I had made the previous summer to Pripyet and its surrounding area. At the time I was travelling through the former Soviet Union in order to research a documentary film. Seven hundred people live within the 30-kilometre exclusion zone around Chernobyl. The area is officially uninhabitable, a fenced-off desolate landscape of abandoned farmland. At the entrance to the 'dead zone' I was introduced to a couple of women in their early sixties who had been evacuated in 1986 in the wake of the accident and then relocated to flats in Kiev. A year later, the two women returned to the exclusion zone, cut their way through the fence and went back to their cottages in the woods. At first the authorities tried a number of techniques to get rid of them. For instance, radiologists showed the women various readings taken from the contaminated vegetables sprouting in their gardens. But the women refused to leave, and finally the authorities saw no option but to let them stay. 'It is better to live a day as a nickel than a year as a crow,' the women replied when I asked if they were aware of the dangers. Of course they were aware. They had been told many times about the strontium in their marrows, the caesium in their potatoes, yet they continued to grow them and to eat them. Of course they understood the risks. Was it normal for a pine forest to be rust-coloured? Even so, nothing would stop them sending produce to their family and friends in Kiev. In the summer their grandchildren came to spend a holiday in the irradiated landscape. For the rest of the year the two women were employed at one of the experimental farms inside the exclusion zone. As they spoke cheerfully of their decision to live with radioactivity, it seemed to me that the two women were ultimately nuclear realists. It was shortly after our conversation that I resolved to draw up a map of the atomic wasteland.

I had met other people in the former Soviet Union who simply shrugged their shoulders when I asked about radioactivity. The world is radioactive, they said. Natural background radiation is everywhere. The Brazilian coffee on our tables is radioactive, as is the granite in our homes. There might be radon gas accumulating in your basement, the person you sleep next to has radioactive potassium in them. There is americium in the smoke detector, thorium in the camping gas light, radium in the luminous dial of an old watch, tritium in the dial of the trim telephone. The sheer number of atomic tests over the last fifty years has ensured that everybody on the planet has small yet detectable amounts of strontium, plutonium and caesium in their bodies. Moreover, well-publicized nuclear disasters, such as the fire at Windscale (now called Sellafield) in 1957, the accidents at Chelyabinsk in Russia between 1949 and 1967 and at Three Mile Island in 1979, as well as many other leakages of radioactive material, have contributed to growing public awareness of the incomparable threat posed to health and safety by even small amounts of nuclear waste.

On the way to Chernobyl I took a train from Moscow to Kiev where I transferred to a 'non-contaminated' minibus. In the town itself we changed vehicles again, crossing the dead zone in a suitably contaminated black limousine. The switch is necessarily routine, although our return journey was held up by problems at 'Checkpoint Charlie', the main security gate whose nickname is perhaps another hollow monument to the Cold War or, at any rate, to the incongruity of closed societies in the nuclear age. In this book I have tried to explore the conflict between nature's openness and national security as it affects different countries in different ways. As a result, it has sometimes been necessary for me to draw unlikely comparisons and to make the occasional cross-reference between Chelyabinsk, say, and Hanford, its obvious counterpart in the north-west of the United States. Here lies a Cold War symmetry. Nobody has yet figured out what to do with the very first contaminated material produced by the Manhattan Project in 1943. Most of it continues to be stored at Hanford in underground tanks that are gradually eroding, while their contents remain as hot as ever, both thermally and radioactively.

As one of the nation's main active dump sites, Hanford stores not only its own waste, but also exotic radioactive material from all over the United States. By the time a nuclear submarine is retired, for example, the hull is so radioactive that the middle third must be chopped out and transported to Hanford for burial. In 1990 the nuclear site received a consignment of 828 beagle carcasses, along with more than 17 tonnes of their urine and faeces – relics of a 25-year experiment in radiation at several universities on the West Coast. The Department of Energy has grudgingly yielded to public pressure to remedy the situation, and it now appears that for some time to come the biggest budgets in Hanford will be dedicated to clean-up. As one state representative used to tell his constituents, the future offers many 'waste opportunities'.

PART ONE

1. Atomic City

At 5.20 p.m., Monday to Friday, a silver Ford Taurus pulls off the highway outside Richland in south-east Washington and turns into a disused gravel pit. A middle-aged man with spectacles, dressed in a suit, white shirt and tie, parks the car and gets out. He takes off his jacket, folds it on the back seat and opens the trunk. Out of a flat wooden case he pulls a semi-automatic rifle, snaps a magazine into the gun's breech, then walks to the usual spot with a box of cartridges. He places earmuffs on his head, gets an empty oil drum in his sights, braces himself and opens fire. Bullets rip through the drum in a deafening sequence and thud into the earth behind. After a minute the cartridge box is empty and the targets are torn to shreds. On the ground lies a pile of smoking shell casings. The marksman returns to the car, puts on his jacket and drives home for dinner. On his way out he passes other cars as they enter the gravel pit.

I stared at a couple of the oil drums in the home of Philip Harding, Richland's only native artist. In spite of having been punctured hundreds of times, the drums were still able to carry the weight of his TV and video recorder. The metal was twisted, curled, scorched, rusted. At the last elections in Benton County the Democrat, whilst running for office as the local representative, tentatively proposed curtailing the use of semi-automatics to five days a week. There was an immediate public outcry, resulting in the Gun Club of America pumping so much money into the candidate's opponent, that he didn't stand a chance of being elected.

On the eve of the Second World War, Hanford and Richland were poor farming communities on the bank of the Columbia river, in the badlands of Washington State. The total population was under 2,000. Today, however, 'atomic city' looks like the small-town America of Ronald Reagan's dream, with leafy streets, well-mown lawns and driveways graced with Chevys and Fords. Richland is now wealthy, white and middle class. The inaptness of its name has disappeared. It has a couple of main streets with the usual Denny's, Henry's and Shari's coffee shops and restaurants advertising 'Super Soups and Salad Bar' in giant lettering. The federal building in the middle of town has recently been furnished with the latest state-of-the-art bomb deflectors at the front of the building, in the wake of the Oklahoma bombing. At the weekend people do sport, ride Harleys, play bridge and enjoy the right to carry firearms. There are fifteen golf courses in Richland, some of which are incorporated into the luxury estates that are now being built to the north of town. When you buy a house up there you get a free caddie cart as part of the deal. On Sundays you go to church – no excuses, because there is a church for everyone. On Stevens Street, for example, you can take your pick from the Lutheran Church, the Good Shepherd Lutheran Church, the Bethel Church for the Community, the Baptist Church and the Church of Jesus Christ of the Latter Day Saints. Richland is quintessential Middle America; it is Average-ville. Everybody in the town knows everybody else, and people are suspicious of new faces. The crime-rate is low.

In fact Richland is anything but average. The town boasts proudly that it has one of the highest IQs per capita in the United States. It also claims to have been home to Harold McClusky, the world's most radioactive man. (Seriously contaminated in a nuclear accident in 1976, McClusky was guaranteed to set off Geiger counters at a range of a couple of metres.) Richland is also one of the most security-conscious towns in the world. Its residents go shopping with their security clearances hanging around their necks. Most have a 'Q' clearance, which means that they work for the Government in one way or another. Without exception, they have had their past life screened for unlawful acts, suspect politics and for any other personal traits that might be regarded as a security problem. Jim Acord, an artist from Seattle, and his wife Margaret, had been living in Richland for only a couple of days when two security agents knocked on their door and asked them about their new neighbours. Were they quiet? Did they ever argue? Had they ever seen them drunk? Were there any complaints at all? There are no homeless people in Richland, no alcoholics, no drug addicts, no pickpockets. The town has been

it is still regarded as one of the dirtiest

lain separating Rattlesnake Mountain
as selected as the ideal place for the
cture plutonium for the atomic bomb.
. There was plenty of water from the
tor coolant. There was abundant elec-
Dam. It was flat and expansive, measur-
in km and apart from the small farming
d White Bluffs, the plain was sparsely
ure. The US Government immediately
ord's 1,500 residents were evicted. All that
are tree stumps from the apple orchards,
l (both derelict). Richland, 50 km to the
y town for the sudden influx of scientists
and engineers working on the site. Three hundred thousand people are estimated to have worked at Hanford at some stage during the war. Many of them viewed it as an inhospitable part of the country, but they put up with the high-speed winds and the dust-storms, and just kept quiet about the work they were doing. Security was so tight that only a handful of people knew why the industrial complex was being built in the first place. Rumours circulated about the end product, and varied from ice cream for the G.I.s in Europe to nylon stockings. What was especially puzzling to the uninformed was the huge quantity of equipment and manpower being shipped into the site, while there was no evidence of any product leaving. In fact what they were producing was plutonium, which finally left Hanford in a suitcase in early 1945. It was guarded by two armed but anonymous-looking men, who took a night train to Sante Fe in New Mexico, where the suitcase was handed over and driven to Los Alamos. Carl Gammetsfelder, who later became deputy head of Health Physics at Hanford, recalls a visit his mother-in-law made to the town. 'I remember she caught the train back to Tennessee after staying with us and got home the same day they published the news about the bombing of Hiroshima. She rang to say she'd got home safe and asked if we'd heard the news. She had absolutely no idea

where she'd just been.' Carl remembers the day being a Saturday. It was the first free Saturday that his colleagues had had in months. 'I was elated when I heard the news. We packed the trunk of the car with beer and drove to the top of Rattlesnake Mountain, from where you could see the whole of the Hanford site. We sat up there all day slowly drinking, happy it was over and proud of what we'd done.'

The last fifty years have left a distinct mark on the community. According to the Environmental Protection Agency, Hanford is home to some 1,300 contaminated sites. Yet it seems that Richland has few nuclear sceptics. The Tri-Cities telephone book (including the neighbouring towns of Pasco and Kennewick) lists about a dozen small businesses that call themselves 'Atomic', even though some of the names are unintentionally ominous – Atomic Foods, say, and Atomic Health Center. Children play in the Leslie Groves Park. Families shop at the Atomic Food Market and wash clothes at the Atomic Laundry. Richland High School's football team, known as the 'Bombers', has a mushroom cloud painted on its stadium. Their helmets get an A-bomb sticker for every touchdown. The school's entrance proudly announces it to be the 'Home of the Bombers'. Bumper stickers urge you to 'KNOW NUKES' and to 'BE A NUCLEAR FAMILY, RADIATE LOVE'. The youth group at Christ the King Parish raises funds by selling T-shirts in fluorescent chartreuse, with 'WE GLOW FOR GOD' around the radiation symbol. Painted on the wall outside the Central United Protestant Church is a family ascending to heaven surrounded by atoms. 'We ended the Second World War, and we prevented a nuclear war,' Mike Fox, one of Hanford's nuclear engineers, tells me with pride. 'A more satisfying way to end the Cold War would have been to crush that tyranny.' He is lost for a second in his hawkish thoughts.

Notwithstanding the stronghold of old-timers in Hanford who believe that the public has a lot to thank them for, a sense of pride among the community is beginning to wane. It is part of a national movement, a collective gripe. Over the last fifteen years the American attitude towards the Cold War has shifted gradually as the truth has been revealed about the financial, environmental and health costs that the phoney war exacted on its own people. Communities like

Richland are coming under fire as it becomes clear just how much of America has been contaminated. People living downwind or downriver from Hanford, or from any other Department of Energy site, are now demanding compensation for the years in which they acted unwittingly as guinea-pigs. Radioactivity, so welcome in the 1950s, is now a pariah. Instead of being celebrated for its role in defending the free world, Hanford is now labelled a polluter. 'We've let our critics define who we are,' argues Mike Fox angrily. 'It's the garbage published by the environmental movement and the media that is primarily to blame.' This has created an atmosphere of embattlement in the town. For the fiftieth anniversary of the atomic bomb, Richland conspicuously kept silent. There wasn't a single parade or speech or flag being waved. When Jim Acord sent 100 paper lanterns floating down the Columbia river in memory of the Japanese lives lost, he encountered nothing more than the odd heckle from high school children. Atoms are being scrubbed off the sides of buildings and unpicked from shirtsleeves. Atomic Lanes Bowling Stadium has been renamed the Fiesta Bowl, after the 1930s household pottery that contained uranium. They've painted brightly coloured parrots over the once-familiar atomic motif. The town recently held a competition to re-design the logo for its municipal vehicles and uniforms, something that did not incorporate atoms or a mushroom cloud. The design that won depicted a new dawn: a sun rising over distant hills. The local police now patrol the quiet neighbourhoods with the 'new dawn' logo on the sides of their squad cars.

At 7 a.m. on a weekday Route 4, heading north from Richland, is invariably at a standstill. A string of vehicles a mile long and two lanes wide slows down to pass through the security gates at the Wye Barricade en route to the site's most contaminated spots, known as the '100' and '200' areas. The security guards are required to touch each security pass to make sure it's not a fake. In 1994 the Hanford site employed nearly twice as many people, 18,856, as it did during its busiest plutonium production phase in the mid-1950s (11,000). Five years earlier a commitment was made to restore the site as close as possible to its original condition, and the result of that commitment, the Tri-Partite Agreement, has become a blueprint for Hanford's future. The unholy alliance of the bomb makers (the Department of Energy), the regulators (the Environmental Protection Agency), not forgetting the state of Washington itself, had a vision. The vision involved making the site 'an area open to a variety of public and private industrial, educational, agricultural and recreational uses. A safe and healthy environment where plants and wildlife flourish, and a role model for cost-effective environmental restoration and waste management.' Its sponsors have given themselves thirty years in which to return the Hanford Nuclear Reservation to a Garden of Eden – and so Hanford is now in what the press relations department calls 'full-scale clean-up mode'. Nowadays company brochures are illustrated with healthy picturesque scenes of deer grazing and fishermen casting rods only metres from a reactor. At the entrance to the site, billboards depict scenes of children picking flowers and water-skiers on the Columbia river. Six years after the Tri-Partite Agreement, the job of cleaning up Hanford has become the largest civil works project in the history of the world, as well as the most controversial.

Of all the Cold War military–industrial sites throughout the United States, Hanford is the most radioactively contaminated or 'crapped up', as the jargon has it. The site is estimated to hold two thirds of the Department of Energy's high-level radioactive waste. It has 1,377 waste sites in the form of basins, pits, tanks, ponds, trenches and cribs into which everything from high-level liquid waste to contaminated rubber gloves has been thrown. There may be as much as 190 kg of plutonium buried in the soil, and 4.4 tonnes of plutonium in metals, solutions, scrap in glove boxes, vaults, piping, tanks, ventilation ducts and office areas. A third of the site is contaminated with tritium and a plume of carbon tetrachloride extends over 18 sq. km. The records of what waste is where on the site are incomplete and patchy. In the 1950s and 1960s, radioactive waste was about as unimportant as any other waste; if it got in the way, you buried it, burnt it, threw it in the sea or simply moved elsewhere. But as the rest of the country began to comply with new environmental and health and safety regulations, the Department of Energy (DoE), remained unanswerable and untouched, owing to the Atomic Energy Act. It was agreed that places like Hanford had the nation's security to think about. Eco-regulations simply got in the way of a priority. As soon as the Cold War ended, however, the sites were forced, kicking and screaming, into compliance with the regulations. What the regulators found on inspecting Hanford for the first time was a mess. 'I don't think anyone understood what we were dealing with when the Tri-Partite Agreement was signed,' says Mike Berrichoa, head of the press department for Westinghouse at Hanford. 'I remember the press conference. At the time I was working as a reporter for local radio and recall the speakers, full of optimism, talking about returning the site to its rightful condition. I got up and asked, "How clean is clean?" No one had an answer, it kinda put them on the spot.'

On a wide bend of the Columbia river, at the northernmost point of the site, nine silent nuclear reactors stand in regimental fashion along the riverbank. Each is a cluster of large, grey, concrete blocks with a single stack: the reactor's chimney. This is the '100' area. None of the reactors has names. They are known only by the random letter assigned to them. There is a 'B' reactor, an 'F' reactor, a 'C' and an 'H'. Balls of tumbleweed blown across the vast open spaces gather against their security fences, caught by barbed wire, somehow emphasizing the fact that the reactors are now disused, unwanted and empty. The security huts are permanently locked. The original 1950s Atoms Café sign at 'K' reactor hangs loosely off its frame. Three hundred and fifty metres from the river are the 'K' basins, where over 2,000 tonnes of spent fuel has been left unreprocessed. The spent fuel rods are stored in 4.5 million litres of water at a depth of 14 metres. If somebody was to pull up a rod from one of the basins, by the time it broke the surface of the water, he or she would be dead from radiation poisoning. The spent fuel is from the 'N' reactor, the last operating reactor on the site. 'N' was opened by President Kennedy in 1963, only weeks before he was shot in Dallas. It was prematurely closed in 1987, following the Chernobyl accident, due to a heated public debate focusing on the design features it had in common with the Russian reactor. The spent fuel rods, some of which are twenty years old, are corroded and leak a cocktail of plutonium, strontium, caesium and tritium into the storage basins. Their forty-year-old concrete basins are cracked and the contaminated water is seeping into the soil. The main priority at Hanford now is to remove the fuel rods from the storage basins, clean out the sludge at the bottom and halt the movement of the radionuclides to the river. Any significant leak from the basins into the river would spell disaster, not least because the Columbia river supplies the southern part of Washington State and Oregon with fresh water and fish, and also irrigates thousands of square kilometres of fertile farmland. If it became known that Washington's wheat and fruit were irrigated by radioactive water, the farming economy would crash. Actually, the water in the Columbia river is already contaminated; strontium, carbon tetrachloride and tritium have been leaking into the river for some time, but, once diluted, the amounts are not considered by the experts to be a risk to public health. 'One big worry is that we might have an earthquake large enough to crack one of those basins open,' Berrichoa told me. The area has a minor earthquake almost once a month. The last time I was there, in September 1995, I was sitting in a coffee shop on George Washington Way when a tremor measuring 2.4 on the Richter scale made the coffee jump out of my cup. For many years it was thought that the Plutonium Finishing Plant would collapse if a major earthquake occurred. This would result in the dispersal of contamination across hundreds of kilometres.

Flat-roofed, anonymous chemical-separation plants litter the Hanford site like ungainly battleships. These concrete buildings accommodate the miles and miles of piping needed to separate plutonium from the uranium after it has been 'cooked' in the reactors. The highly radioactive acidic soup, which is produced by separating the elements, is widely thought to pose the greatest environmental and safety danger inside the DoE complex. Two hundred and seventy-five million litres of this particular liquid waste are stored in 177 giant steel tanks buried in the '200' area. Built in the 1940s and 1950s the three-storey-high tanks were only supposed to be used as temporary storage, for a maximum of twenty years. Sixty-seven of them are known to have leaked around 4.5 million litres of chemical sludge including radioactive strontium, caesium and tritium into the soil. It is anticipated that a new leak will be discovered each year. 'No one really knows what is in the tanks,' says Mike Berrichoa. 'Some of the contents are the consistency of peanut butter. Some are like motor oil or thinner, and some have evaporated over the years and formed into crystals. We are now starting a sampling programme to find out exactly what we are dealing with.' Hanford's most infamous tank, 101 SY, is known as the 'burping' tank, because it often releases a gigantic belch of highly flammable hydrogen gas. A vast cauldron of chemical reactions, it reaches temperatures well over boiling point. By means of a video camera inserted into the tank, workers can watch the contents erupt with such force that the solid crust that forms on the liquid surface is broken. Environmentalists feel that the gas might one day ignite, causing an explosion similar to the one at Kyshtym in Russia in 1957, where a tank exploded scattering highly radioactive waste over hundreds of square kilometres. In the meantime, however, millions of dollars have been spent studying the tank. The problem has at last been solved by stirring the contents slowly with a giant eight-tonne spatula.

In spite of the ambitious hopes of the Tri-Partite plans, not much in the way of clean-up has occurred at Hanford over the last six years. Westinghouse, the site's main contractor, argues that the work to date has been concerned with assessing the scale of the problem and attempting to stabilize the more serious hazards. But with the total cost of cleaning up Hanford estimated as high as US$50 billion, the taxpayers are understandably displeased. Since 1989 they have witnessed nearly US$2 billion a year being spent on a colossal amount of paperwork. One of the main complaints is that the employment structure is top heavy. A 1994 government report gave the example of the waste tanks employing ninety-three workers as opposed to 146 managers. As the clean-up money started to roll in, the workforce at Hanford seemed to enjoy a honeymoon period; in any case, they were relieved that they still had jobs to go to in spite of the end of the Cold War. More golf-courses, a new stadium and a shopping centre were built in and around the town. This 'business as usual' attitude caused a lot of anger with those who wanted to see more tangible results at the site. The project has been accused of floundering in a legal and regulatory morass and of having little to show for the levels of expenditure. Not surprisingly, this has intensified the atmosphere of anger and distrust and turned Hanford into a battleground between Congress, the three parties of the agreement and the public. Congress has retaliated by cutting Hanford's budget, as well as threatening the Department of Energy with closure. This combined pressure is having some effect. Priority is now being given to the most serious clean-up problems. Employees are being made redundant and contractors, like Westinghouse, are being employed on performance-based contracts. 'Perhaps it is true to say that we weren't as careful with the budget as we could have been,' says Mike Berrichoa. 'But it's a whole different world dealing with radioactive waste. We simply did not know what we had out there so we had to be prepared for anything. You take state regulations, federal regulations and federal law and Department of Energy's orders, on top of which you've got company policy and procedure, you hand all that to a worker and the work becomes a by-product of the process.'

In Richland there is a syndrome known as 'becoming Hanfordized'. A person who ceases to notice the thousands of tax dollars being poured down the drain, or who turns a blind eye to incompetence is said to have 'become Hanfordized'. Gary Lekvold swears that as long as he lives he will never 'become Hanfordized'. 'Things don't get done here in the same way that they do in the rest of the country. Something that would take a week normally, would take six weeks here and cost thousands of dollars more. Here there are layers of bureaucracy and incompetence that you don't get outside.' But outside is exactly where Gary Lekvold has been ever since he blew the whistle on the Westinghouse Corporation after witnessing a number of security violations at Hanford. 'We have a culture here with top levels of secrecy that has existed for many years,' explains Mike Berrichoa. 'It's a system that promotes internally, so you get second and third generation workers coming up through the ranks. This means that everything that is customary is perpetuated. These people who "blow the whistle" stand out in the crowd because they have a personality that is not mainstream, either that or they do it for political or financial benefits.' Inez Austin says that as soon as she became known as a whistleblower, her neighbours and friends in Richland stopped calling. Even her daughter's school friends stayed away. 'I used to be a pretty popular person in this neighbourhood,' she told me, her arm gesturing towards the street outside her window. 'Then overnight no one talks to me. It's like being a leper.'

There is a Saturn car advertisement on American television that tells the story of a black worker who stops the assembly line because he notices that a small fitting is missing from one of the cars. All the workers get on their hands and knees to look for it. The fitting is eventually found and returned to its rightful place. The factory foreman comes out, all smiles, to congratulate the worker who spotted the mistake, and the assembly line starts up again. The message is obvious: Saturn regards its whistleblowers as a necessary requirement of the workforce. Thanks to them Saturn cars are that little bit special.

The whistleblower is now an acknowledged component of most industrial workforces. Corporations today have departments for employee concerns and internal and external complaint procedures to resolve problems at work. This hasn't always been the case. It is the experiences of the nuclear industry's whistleblowers that has helped pioneer the rights of the workforce. The case of Karen Silkwood is now well known. In 1974 Silkwood, an employee with the nuclear power company Kerr McGee, died in a mysterious car accident in Oklahoma City while on her way to meet a journalist from *The New York Times* with documents allegedly pertaining to illegal practices at the plant. The Department of Energy, or the Atomic Energy Commission, as it was called, is notoriously secretive and isolated. The government agency, principally responsible for the manufacture of nuclear weapons throughout the Cold War, has been protected from birth by a number of acts of Congress that ensure issues of national security continue unhindered. Out of that culture was born the nuclear power industry. It too has been protected and nurtured by successive governments throughout its life. In the States, the sudden and massive expansion in the 1960s and 1970s of this new industry gave rise to a number of problems: technical problems, security problems, health and safety problems and employment problems. Many of these critical areas were overlooked by the nuclear power corporations. They, after all, were riding on the pig's back, making good money, so why spoil a beautiful future? Out of this dark and secretive culture ventured the nuclear whistleblowers, one by one, into the glare of public scrutiny.

The first round of Hanford whistleblowing started in 1986 with an employee by the name of Casey Ruud, who alleged that Rockwell, the main site contractor, was ignoring environment laws, audits and safety conditions. Ruud became famous when he testified in Congress, at a hearing that contributed to Rockwell losing its contract. In the years following the Ruud case, however, more and more whistleblowers stepped forward, as Hanford began to undergo its painful transition from bomb-making to clean-up. A clash of cultures between the old and the new workforce, not to mention changing political attitudes and increasing pressures from Congress, environmentalists and journalists, served to burst the bubble. The Department of Energy began publicly to condemn its contractors' retaliation upon whistleblowers and gradually installed a number of employee concern programmes. Eventually, in 1992, Congress amended the law in order to protect whistleblowers by allowing them to file a complaint with the Department of Labor. If they won their case, they could be reinstated and receive full back-pay. The barrage of media interest transformed the whistleblowers into heroes and propelled ordinary engineers and scientists, like the Jack Lemmon character in *The China Syndrome*, momentarily to stardom. They were flown to Washington, filmed and photographed with the Secretary of Energy and praised publicly for their courage in speaking out. But as soon as the show was over, they were left to pick up the pieces and go home, their careers in tatters and their domestic and social lives ruined.

In 1990 Inez Austin worked as a pump engineer on the liquid-waste tanks in the '200' area. She was asked by a manager to sign a report authorizing the use of a certain process, to pump liquid waste from one tank to another. Inez refused to sign as she was concerned that the proposed technique was not applicable and was potentially dangerous. The manager began to hound her, ordering her to sign, but Inez continued to refuse, so he fired her. Later she was reinstated. At which point she says the 'purgatory started'.

'The first thing they told me was to go see the psychiatrist. Well, I refused. Then rumours started to circulate the office that anyone caught talking to me would be fired.' The work that she was doing ground to a halt and her office was moved outside to a trailer. Throughout the winter, the heating in the trailer unexpectedly fluctuated from being too hot one minute to freezing the next. Then the trailer was mysteriously removed, so Inez worked all day doing odds and ends from her car parked in the parking lot. At home she began to receive threatening phone calls. One caller, she says, threatened to cut off her granddaughter's head and use it as a fruit bowl. Her house was broken into on a number of occasions; doors and windows were left open, but nothing was ever stolen. Her mail was not delivered for half a year. The pressure became so bad that her son changed his name and her daughter moved out, while Inez's mental health deteriorated to such an extent that she became resigned to the fact that they were going to kill her. 'I didn't think that they had the power to do things like that,' she said, as she curled up on her sofa. She has attempted to look for other jobs, but claims that she has been blacklisted and is therefore unable to find employment at any DoE.

Inez's story is not unusual. There is a small group of whistleblowers who still live in and around Richland, unable to leave the scene of the crime. They lead quiet, lonely lives ostracized by their communities. Life for Inez is marginally better these days, the threatening calls and break-ins have stopped and her daughter has moved back in. Despite the fact that no one talks to her, she still punches in at work every morning and is paid US$52,000 a year to work as a skilled pump engineer. She has her office back and keeps herself busy most of the day by organizing office parties and doing administrative work.

'Thyroid problems? I've got thyroid problems. Everybody has thyroid problems around here!' the wife of one of Richland's chief health physicists told me without apparent irony.

Brenda Weaver keeps her daughter's eyes in a small cardboard box. 'I don't know why I collect them really. Sort of sentimental, I guess.' Her finger lovingly fishes inside the box, turning each eye over, as if she was looking for one in particular. The beautiful painted glass balls record the growing of Brenda's daughter, Jamie. There are ten or twelve sets of varying sizes. They stare at you coldly. Jamie was born in 1965, without eyeballs in her sockets. Brenda Weaver has been taking lambs' blood pills for hypothyroidism for most of her life. A year ago her doctor told her that her thyroid glands were no longer functioning, and that the most likely cause was exposure to radioactive iodine-131, from the Hanford plant. The diagnosis confirmed Brenda's fears.

After the Second World War her father, like many other veterans, was offered land north of Hanford at a competitive price. The initiative was aimed at giving war heroes a fresh start. Although the land was dry, hard and inhospitable, Brenda's father accepted the offer and got to work planting crops and irrigating. Brenda, who was born shortly afterwards, has happy memories of growing up on the farm. With her sister and two brothers, she spent most of her time outside playing in the dirt or swimming in the Columbia river. They ate the produce from the farm and drank the milk from her father's cows. She remembers being proud of the 'atomic city'; it was there to protect them. So they welcomed the men from Hanford on the four or five occasions each year when they visited the farm. As a rule, two people would show up, dressed in white suits and carrying various types of equipment. Brenda would watch as they collected samples of earth and plants from the area and checked the dead fish on the river-banks. By the mid-1950s, she remembers her family getting sick a lot. Her sister always had a bad cold and often contracted bronchitis or pneumonia during the summer months. At the age of fourteen, Brenda had an operation to take out a ruptured ovary, and her sister was diagnosed as having appendicitis. The following year during the lambing season, an extraordinarily high number of lambs in the area were born deformed. The incident became known by the local farmers as 'the night of the little demons'. Some newborn lambs were unrecognizable lumps, others had two heads, three legs or no eyes. At the age of eighteen, Brenda gave birth to Jamie. 'They didn't bring her in right away. I was pretty drugged up because it had been a difficult delivery, but it seemed like I was the last to know. When they did bring her in, I just cried. I didn't know anything about blind people.' One morning in 1986, twenty-one years later, Brenda opened her local paper and read an article which connected the night of the little demons with airborne emissions from the Hanford site. 'It hit me like a blinding light. I realized that it was me who had been blind. Suddenly everything made sense to me: the dead fish, the animals, my sister's health, the warm pools in the river that we swam in. I knew instantly that Hanford was to blame for my baby being born with no eyes and for my own health problems.'

Glade Road runs north across the high plains between Pasco and Moses Lake. In the last few years it has been renamed 'death mile' by local farming families, on account of the disproportionately high cancer rate. Betty Perks lives just off Glade Road, not far from Brenda Weaver's farm. Betty, also the daughter of a war veteran, is convinced that they were sold the land because they were patriotic and therefore less likely to complain about the Hanford site. She remembers a small aeroplane regularly flying low over their farm in the early 1960s. The plane carried a large black box beneath its wings and appeared to be collecting samples. Her children often waved at the pilot as he flew past, but she remembers that he never waved back. She recalls the weeks when her husband would come in after working outside all day, with dust and wheat husks in his hair and weeping sores on his scalp. She remembers men from Hanford coming to the farm and asking what they ate and drank. She's heard stories about the local farmer whose dead cow was removed one night and put back the following day in exactly the same place and position, with its internal organs removed. 'Foxes or dogs,' his neighbours told him, but there was no blood and the skin was cut with surgical precision. Five out of seven members of Betty's family are taking medication for hypothyroidism. Her fourth child lived for only a day with underdeveloped lungs and a bleeding cranium. Her daughters have had surgery on their thyroid glands, leaving a scar on the neck known locally as the 'Hanford necklace'.

Further north, in the small town of Ritzville, Laverne Kautz takes me down three blocks of her street. It's a quiet neighbourhood of neat bungalows with carefully manicured front lawns, net curtains and cuddly toys placed in the windows. 'They've just had a death in the family,' she says, pointing to a house as we pass. 'The father of that family has just learned that he has cancer,' she says, referring to another. 'Their son died of leukaemia, and the woman who lives here has cancer and she's only in her mid-forties.' On and on it went until we reached the end of the street. In the thirty years preceding the Manhattan Project the death rate from cancer in Ritzville, she says, was just over two a year. In 1990 seventeen residents of Ritzville died from cancer. Laverne recently learned from a dose reconstruction study that Ritzville was a 'red area', meaning that the town exceeded the federal exposure limits. The news did not shock her, it only confirmed what she knew to be true. She considers herself lucky to be alive. Her doctor told her recently that her immune system was so defective that anyone would think she had AIDS.

For thousands of Hanford downwinders the research studies being undertaken to discover whether or not they have been affected by the discharges from Hanford are simply an insult or a way of postponing compensation payments. They *know* they are affected, because they live with it day to day. Brenda Weaver, Laverne Kautz and Betty Perks all knew it the day they opened the local paper ten years ago. They have found their answer; it was staring them in the face for so many years and they couldn't see it. They don't need a scientist to prove or disprove what they now know to be true. 'After seeing the statistics and watching so many loved ones suffer and die from cancer, I have reached my own conclusions,' says Laverne.

Mike Fox is a nuclear engineer with thirty years' experience. He has lived in Richland for twenty-two of those years and states categorically that the downwinders are mistaken. Fox has done his own epidemiological studies. 'Oh I sympathize with them,' he says, 'but frankly their problems have little or nothing to do with radiation. It's what I call victimology, it's an epidemic that's sweeping America. Our constitution talks about individual's rights, but somehow that has been translated to mean group rights, whether it be women, gays and lesbians or downwinders. I've been telling my friends to develop some illness,' he jokes, 'because the downwinders are going to get compensation, there's no doubt about it.'

He often refers to a survey that looks at the perceived risks from radiation as opposed to the real risks. The correlation between the two sets of figures, he tells me with glee, is 4 per cent. 'There are a number of reasons for this exploitation of risk. They are as follows,' he begins in a well-rehearsed tone of voice. The 'greens' are top of his list for blame. A close second is the media, and to prove it he quotes an editor of the *Washington Post* as saying, 'We don't print the truth, we don't pretend to, we print what people tell us.' Third on his hit list are the politicians and last the consultant engineers, 'scientific whores, who know that exploiting the public's fear means more money.'

Mike Fox is well known in Richland. His uncompromising language identifies with a strong desire to regain control of the country, a mixture of Gingrichism perhaps and a militia-movement tendency. He believes that the success of America as an industrial power has been achieved through the advancement of science and technology. He mourns the fact that 80 per cent of Americans are not aware of the existence of natural background radiation, that 50 to 60 per cent of the public don't know anyone with a degree and only 5 per cent of his school friends are what he calls 'scientifically literate'. The paucity of scientific education, he argues, explains why America is now losing its foothold in the world. Part of the problem, he says, is that the environmental regulators have got America's industry 'by the balls'. His campaign to prise the future of the Hanford site from the grasp of bureaucrats and politicians, and to put it back into the hands of the ordinary man, is a source of anxiety in the public relations department. He tells me incredulously, 'The US spends about US$115 billion each year complying with environmental regulations. There is little more than a paint shed out on the Hanford site that has cost one million dollars and has taken six years to tear down. Following the regulation guidelines, we had to determine whether the building we wished to remove contained toxins! Well, we knew it did, it's a paint shed, for Chrissake! But following instructions, we had to take samples of every building material in the shed: the floor, the concrete structures, the roof and the soil underneath the building. All those samples then had to be analysed, which alone costs a quarter of million dollars. Regulations are ruining America. Yet if we ignore these requirements, we run the risk of prosecution.' Fox runs another statistic by me, just to make sure his point is clear. 'Wolf Creek nuclear power station employs 1,200 people, whereas Sizewell B reactor employs 350. My point?' he says, with perfect dramatic timing. 'They're the same god-damn reactor.'

He rose to fame during the 1988 state elections, because much of the political argument centred on whether Hanford was a suitable location for a permanent waste repository. Sick and tired of what he calls the 'outlandish' remarks being made by various politicians, Fox organized a Hanford support group. The 'Hanford Family', as they called themselves, used the same tactics as the environmental movement, their opponents. Three thousand residents of the Richland area, under Fox's supervision, staged a demonstration on Cable Bridge. The event caught the attention of the media and, for a while, Mike Fox became the angry, outspoken face of the Hanford nuclear site. 'We took the initiative to do something that the nuclear industry has never done – fight back. We publicly rejected the rubbish that was being stated about Hanford being a health hazard to the north-west. We explained to the public for the first time, using words of one syllable, what the nuclear industry is about. I am proud to say that we got our point across.' To prove it, he cites the fact that 80 per cent of Benton county voted Republican that year, and although the Hanford Family has disbanded, Mike continues to fight on. The Department of Energy, he believes, is now run by people who are left wing and anti-nuclear, and who have no wish to enlighten the public. Mike confesses to being 'dangerous' in the possession of statistics and admits to enjoying nothing more than destroying the environmentalist case in lectures at universities and conferences. 'A lot of the time they will refuse to talk if they know I am going to be there.'

Just now Mike Fox is carrying out a series of experiments on new isotopes found in the nuclear waste at the site. He is testing their abilities to fight cancer. He says that he has so far witnessed remarkable remission rates for thyroid cancer and leukaemia. As a matter of fact, he got a call recently from the president of Westinghouse, who told him to keep up the good work. 'Someday, he told me, you're going to be a hero,' Fox says with obvious pride. But for many of the people in the pro-nuclear lobby in Richland, he already is.

The Cold War will be a dim recollection by the time the clean-up at Hanford nuclear reservation is over. But even after the property developers move in and litter the river front with golf-courses, tennis courts and suchlike, the water will still be radioactive, and behind the security fencing in the '200' area, the soil will still be contaminated by irretrievably high-level wastes. In other words, no amount of real estate gloss can disguise the fact that Hanford is (and will remain for hundreds of years) 'the dirtiest place in America'.

2. *Tar Baby*

If you drive north on Highway 240 out of Richland at night, you pass a small, floodlit warehouse with the loading bay rolled up, exposing its interior. Sitting at a table in the middle of the warehouse will be a solitary figure reading the paper or listening to the music of the underground band, Throbbing Gristle. Jim Acord is the only person in Richland who listens to Throbbing Gristle for relaxation. He wears a 'Fast Flux Test Facility' T-shirt and thick braces, held down with safety pins, keep his trousers up. The left lens on his only pair of glasses is cracked straight across his field of vision. He drinks Bushmills whiskey for breakfast and sleeps on a blanket rolled out on his studio floor. His staple diet consists of coffee and doughnuts. His laugh, a fruity guffaw, is the result of too many cigarettes and a bad chest cold. Bedtime reading for Jim Acord is *The Life and Art of Joseph Beuys.* Two years ago Acord sold his house in town and moved into the studio because his wife, Margaret, decided that she couldn't take Richland and its people any longer and left him to move to San Francisco. Her parting words to Acord were, 'Be careful of what you wish for.' Now he lives alone with a stray cat named Eddie, who 'works nights' catching mice. The studio overlooking Hanford nuclear reservation is flanked by the highway and the railroad; twenty-four hours a day trucks roll past on one side and nuclear waste trains clatter on the other.

Acord runs the Hanford Sculpture Park, but you needn't rush to visit: there is no park and the only sculpture is a granite sheep's skull that he carved several years ago. Acord's name for his studio refers more to his vision than it does to his present circumstances. He dreams of combining nuclear technology and art – and he has gone some way to making his dream come true. For one thing, he is the only individual in the world to hold a radioactive materials handling licence, and for another, he is the proud owner of a dozen highly radioactive nuclear fuel rods.

It was his vision that brought him to Richland in the first place. For the last fifteen years Acord has been trying to produce a sculpture that's able to encapsulate the duality of art and science. He is zealous, almost evangelical, in making the point. 'We can make a spearhead or a helicopter using a technology, sure, but until we make a statue, it is not really ours,' he told me, as we looked round the sculpture park. 'We've had the Bronze Age, the Iron Age and now we're in the nuclear age. Now is the time for art to make its contribution to society's understanding of this technology.' His idea is to use nuclear technology to create a sculpture that can be exhibited permanently on a nuclear site. In some ways the argument is compelling. Sculpture, according to Acord, is the art of shape – the purposeful and aesthetic arrangement of mass and void – and that shape is the critical property of uranium and plutonium. After all, as he observes, it was the ordered arrangement of the mass and void of these two metals that procured our twin ticket into the nuclear age. Obviously, no more appropriate materials exist for sculpture than those whose essence is altered so dramatically by shape alone.

Unfortunately, the only place that Acord can get his hands on the appropriate technology is at a Department of Energy research laboratory or at a university. The first time he went to Richland, in 1988, was to lecture to art students at the nearby Columbia Basin University. Being up-to-date with the nuclear waste controversy, he was intrigued by the proximity of the Hanford nuclear reservation and, out of simple curiosity, he went to have a look at the place. Immediately he was impressed by the sheer scale and craftsmanship of the technology. It occurred to him that Richland was exactly the sort of community that would understand what he was trying to do. But instead of contacting the nuclear facility in advance to discover if the management would be receptive to his project, he just turned up in Richland with a suitcase in his hand.

With the benefit of hindsight, he now sees that the Hanford nuclear reservation is perhaps the worst place for someone like himself. 'You know, I made two big mistakes when I came here,' he laughs. 'I thought I would find a bunch of liberal scientists in Richland. I'd read all the early history and I knew about Szilard, Oppenheimer, Fermi and all those guys. They would sit around reading poetry, quoting Sanskrit and playing the violin, and I thought, "Wow, man! What a community!" What I found, however, was a community that was right wing, reactionary and Republican. The other mistake I made was to think that such a town would welcome or accept an artist. I had always seen art and science as two parallel paths to the godhead of truth and knowledge, but what I discovered was a terrible discrimination against the arts. Everyone thought they knew what art was; it was the inexpensive hobby for the bored spouse at home or what your least gifted child would go into if he flunked calculus twice in a row.'

Acord's first year in Richland was spent explaining his theory to anyone who would listen. He gave talks to engineering societies and men's fraternal organizations about history of art, its role in society and its relationship with technology. 'In that first year, every show I did flopped. The lights would go on, I'd turn the slide projector off, look out at the audience and every jaw was slack. Every eye was glazed over and half the people were asleep. These people didn't know who Rodin or Michelangelo were. At the end of each show my punch-line would be, "In order for me to utilize your technology to create art, I need your help. I need the help of expertise in nuclear physics, chemistry and engineering. Is there anyone here who is willing to meet with me on an informal basis and answer some of my questions?" I just got more blank looks, a total zero.' I asked why he didn't pack up and leave. 'Well you know, Jerry, it takes a certain tenacity to be a sculptor. You don't do a granite carving without standing there flat-footed for hundreds of hours just working on it and solving the problems. One of my strengths, all along, has been to keep working on something I know is not going well. I keep working and keep working, until it works. That's how things get done.' He leans forward and whispers, 'I will say though, as a personal aside, total zero would have been a good time to quit.' He rolls back laughing and helps himself to another slug of Bushmills.

Acord's fascination with nuclear technology began in 1980. Whilst working in a carving workshop in Barre, Vermont, a town that advertises itself as the 'Granite Capital of the World', he toyed with ways to combine metalwork with stone carving. Aware that all magmatic stones contain traces of metallic elements, he explored the properties of granite and discovered that the metal of greatest interest was uranium. Two years later he returned to his studio in Freemont, Seattle, where he carved tombstones for a living and continued working on the project. He decided to extract the uranium from his own supply of granite. Then, as an artistic statement, he planned to place the element in metal film containers inside a granite stone carving. As he researched the task ahead, he discovered that extracting uranium from granite required not only vats of cyanide but also an advanced understanding of chemical processes. And so, being practical as well as creative, Acord decided to look for a simpler solution.

In 1982 Congress signed the Nuclear Waste Policy Act, and thereby set in motion the widespread hunt for a permanent storage site for nuclear waste. The Hanford reservation became one of the clear favourites. The debate in Seattle, home of many liberals and environmentalists, was particularly loud and shrill. Acord attended some of the angry public hearings and realized that it was an inappropriate moment to start creating his own radioactive waste. Instead, he decided to use the waste generated by the utilities, thus in his own way contributing to a solution of the waste problem. 'I began calling nuclear power plants asking them for small quantities of spent fuel,' he says, laughing at his own naivety. 'Of course, I quickly learned that it was a closed shop. No one wanted to talk to an artist and everyone was petrified of doing anything but walking the straight and narrow.'

In response to the lack of interest shown at his talks in Richland, Acord cut out all references to modern art and changed the presentation to incorporate two slide projectors. He started putting up slides of a nuclear fuel assembly and a twelfth-century French reliquary or the Fast Flux Test Facility reactor and Stonehenge side by side, illustrating the similarities in terms of design. He refined it significantly by putting together a video interspersing footage of his own sculpture and of workers at the site. The combination of video, slides and his life story, culminating in the move to Richland, suddenly began to pay off. A group of nuclear physicists and chemists, including Mike

Fox, volunteered to form a technical advisory team. They now meet one morning a month at Denny's Coffee Shop in the middle of Richland.

'The first thing I realized when we met was that I didn't know enough about their world even to communicate with them,' says Acord. 'I was teaching them about my art, but I knew nothing about their science. To find out what I needed to know I had to talk to them on a higher level, so I went back to school.' Acord, who admits that he flunked maths at school and recalls geometry as being the only technical subject he understood, enrolled at the Tri City's University in Richland and began two courses in advanced nuclear engineering. His classmates, graduates with engineering degrees completing their masters, had a betting pool on whether 'the artist' would pass. The odds were against him. The drop-out rate for the courses was 25 per cent. But they underestimated the commitment of Acord, who worked day and night, seven days a week, and finally managed to pass. 'I knew I had to pass those courses. I was the only non-techy in there and I knew that if my sculpture was to contribute successfully to nuclear technology, I had to understand the process every step of the way. I did it by committing that shit to memory,' he laughs. 'I was like some monk on the Isle of Iona, who is totally illiterate, but has memorized the entire Gospel and can write it down with a quill pen.'

On the table in the centre of Acord's studio, among the papers, ashtray and bottle of whiskey, is a strangely luminous bowl with a piece of dried sagebrush in it. When Acord waved his Geiger counter over the bowl, it clicked excitedly. Fiesta, the Art Deco tableware manufactured in the 1930s and beloved of Andy Warhol, was famous for its Mango Red glaze. Nevertheless, casual collectors may be unaware that the pieces should be treated as hazardous waste, for the simple reason that their vivid colour (more orange than red) is produced by a radioactive material – uranium oxide. After the war, as it became known that Fiesta tableware was indeed rather dangerous to have around the kitchen, people began to throw out the much-loved cups and saucers, and the company went out of business. The orange bowl on the table is the key to Acord's entry into the nuclear age. In

1983 he discovered that pottery was not covered by the Radioactive Materials Handling Act. 'I started going around second-hand stores and flea markets in Seattle and buying every bit of Fiesta I came across,' he recalls. In an attempt to extract the uranium, Acord set about crushing the pottery and panning for the ore using biscuit-tin lids and the sink in his studio. The method was surprisingly effective, but unsurprisingly not very popular with his colleagues. The artists' community in Freemont was worried that he was placing his own health and that of everyone else in the building at risk. They were also concerned that Acord's work was helping to glamorize the nuclear industry. This became such a divisive issue that the Arts Council passed a resolution making Freemont a nuclear-free zone, as a result of which Acord had to move out.

At the same time the Washington State Department of Radiation Control called, wanting to know what the hell an artist was doing working with uranium. Acord went downtown to explain his artistic vision. In spite of his vigorous protest, the Department concluded that he was in fact mining and milling uranium and was therefore required to own a licence. Without the licence, it was illegal for him to own more than fifteen pounds of Fiesta ware at any one time. Acord appealed three times. 'I started teaching art history to the State Department of Radiation Control, I did drawings specifically to explain to them what I was trying to achieve. I showed them sculptures that were over 800 years old that had preserved fingernails and hair. I was trying to get across the idea that putting radioactive waste in a sculpture was just as valid as doing anything else with it.' In other words, Jim Acord was becoming a celebrity. Discussions in the bars and coffee houses of Seattle focused on whether he was a martyr at the hands of the federal bureaucracy or a traitor to the environmental cause. The same debate plagues him to this day.

Having passed the courses in advanced nuclear technology, the conversations with his technical advisory team at Denny's Coffee Shop began to flow, and it wasn't long before the concept for a new sculpture came to him. 'What is unique about the nuclear age is a dream that is at least 3,000 years old: the ability to transmute one element into another, the philosopher's stone, lead to gold and all that. The Romans worked on it, Arab alchemists were working on it in the twelfth century, but it wasn't until fifty years ago, using nuclear technology, that we learned how to do it. I discovered that I could transmute one element into another using fast-neutron capture and make a sculpture out of it.' Acord and his 'Ph.D.s', as he calls his technical advisory team, settled on transmuting element 43, technetium-99, a highly radioactive gamma emitter that has a half-life of 212,000 years into ruthenium-100, element 44, which is radioactive for fifteen seconds. If he could convince the authorities to let him use the Fast Flux Test Facility reactor to transmute the technetium, according to the Ph.D.s he could then take small quantities of the ruthenium and coat it on to a sculpture using electro deposit or plasma flame spraying techniques. 'This is what keeps me here,' he says enthusiastically. 'Despite the dire poverty that I'm living under, this is such a good idea. I can't let go of it.'

Technetium is the first man-made element. Carlo Perrier and Emilio Segre from the Royal University in Palermo created it in 1937 using the 94-centimetre cyclotron at the University of Berkeley, California. Element 43 had always been a great mystery and remained as a blank in the periodic table until the two scientists took element 42, molybdenum, and bombarded it with neutrons to make element 43, calling it technetium because it was technically manufactured. The isotopes were established as being technetium-95 and technetium-97. It wasn't until the following year that Glenn T. Seaborg and Emilio Segre discovered technetium-99. All this was eclipsed by the creation of plutonium at the same university, and the rest is history, as they say. Technetium would have remained a rare lab curiosity if the bomb had not been made. The bombardment of uranium or plutonium results in fission products that fall into the middle of the periodic table. As a result, we now have tons of technetium-99 as a nuclear waste product.

In 1992 something happened to Jim Acord that meant he would never be able to leave Richland as easily as he'd arrived. The 4-metre-long wooden replica of a reactor fuel rod hanging from his studio ceiling is his daily reminder of the fact. At the time the event seemed like a real breakthrough but, looking back, he wishes it had never happened. He gave his slide and video presentation to 200 nuclear engineers at an International Conference on Fast Breeder Reactors in Richland. After the show, six representatives from Siemens in Germany approached him, full of praise, wanting to help in any way they could. Artists like him, they believed, were exactly what the industry needed to create a better image and a greater understanding with the general public. Would he, they asked tentatively, be interested in accepting, as a gift, twelve breeder blanket fuel assembly rods to make sculpture with? Siemens at the time were dismantling a brand new fast breeder reactor that had never been licensed due to political changes in Germany. They agreed to meet the following morning to discuss details.

There were twenty-eight of them for breakfast that morning. Besides Acord and the Siemens team, the proposition alarmed representatives from Westinghouse Hanford, the Department of Energy and the Nuclear Regulatory Commission, who were unsure about the wisdom of giving fuel rods to an 'artist'. Nobody really knew who this guy was, he'd never possessed a security clearance and, for all they knew, he might make a bomb with them. 'The Germans told me that I could have as many fuel rods as I wanted, to make sculpture with. They also told me that I'd need a handling licence because they were radioactive. This is what has got me stuck in this thing, they've been my 'tar baby'. But not knowing any better, of course I said I'd have them, they're beautiful. You know when the Germans make shit, they REALLY make it good. I also thought that this would get my foot in the door. Instead of being the perpetual outsider here, I'd become a member of the club.'

It took Jim Acord a year and a half to get his 'special nuclear materials' licence. He regards it as an achievement so significant that he's had his licence number tattooed on the back of his neck. He is now the only individual on earth to hold one. It is likely to stay that way. Every regulatory loophole and door that he squeezed through was firmly slammed behind him. Copies of his application letters are pinned to every bulletin board in every office at the Nuclear Regulatory Commission, as a reminder of the debate that raged as to whether Jim Acord and his Hanford Sculpture Park was their worst nightmare or exactly what the industry needs. He defied his critics in Richland and, despite their disapproval, he forced his way into 'the club'. He is now officially a member of the nuclear establishment. 'Here I was, this artist who was ready to import a ton of uranium into the country, and everybody was screaming at each other to CUT ME OFF!! But you know, there were just enough people who were prepared to help me to write the proposals and the licence requirements. I had to translate 600 pages of German scientific documents into English. The bottom line was, I met the requirements. I'd done the courses. I knew about radiation protection, I could fill out the forms and I had my technical advisory team.'

As one of only five licensees in Richland, Acord sits on committees for emergency evacuation and radiation safety. He is on the board of directors of the Nuclear Research Medicine working group, he's a member of the 'B' Reactor Museum Association and the American Nuclear Society Public Information Library Committee. He is registered with the International Atomic Energy Authority in Vienna, the Federal Ministry of the Environment in Germany, the Nuclear Regulatory Commission in the United States and the Department of Radiation Control in the state of Washington. But the cost has been dear. As soon as he became the proud owner of twelve assembly rods, his time and overheads spent on the sculpture project tripled overnight. The licence costs him thousands of dollars a year and involves huge liabilities. He's had to form a corporation and is its director, radiation safety officer, material control and accountability custodian, accountant and security officer, all rolled into one. Every month he is required to check his twelve breeder blanket fuel assembly rods, stored on the Hanford site, for radiation leaks with his own tamper-indicating material control and accountability seals and calibrated Geiger counters. Each year the regulations tighten and the costs spiral accordingly. As a result, he is three months behind on the studio rent and is currently living off credit cards, donations from friends and his work on the lecture circuit. 'There's no support system for artists in Richland, everybody says "'Atta boy, Jim," but no one wants to help. Y'know, Leonardo da Vinci used to support his science off his art sales. It just happens that today the pendulum has swung the other way,' he says. 'In the seven years I've been in Richland, not a single person has bought a drawing or sculpture.'

Time is running out for Jim Acord. His hopes of transmuting technetium at Hanford evaporated when the Department of Energy decided to close the Fast Flux Test Facility reactor at the end of 1993. The idea would never have got off the ground anyway. The Department of Energy told him facetiously that he could use the reactor if he raised US$80 million. The last time we met was in a brightly coloured Mexican restaurant in Richland. Jim was depressed. He'd spent all day with his Ph.D.s, and now he was exhausted and despondent. It was becoming clear to him that the artistic vision that had plagued him for fifteen years was going to remain just that – a vision. He told me stories of working nights, driving fork-lift trucks at Tater Boy, the local potato crisp factory, to earn a little extra cash, while delivering lectures all day to nuclear physicists. He has put so much into his sculpture already, that he's spent himself. He's broken up his marriage, ruined his health, lost his money and spends his time in almost total isolation. 'My existence here is such a tightrope act. In one way, I'm interacting with a bunch of unrepentant right-wing scientists and, on the other hand, a lot of my sympathies are with the airhead environmentalists who don't know what is really going on, and here I am trying to keep both sides informed of what I think is the right view on it and I'm tired of it.' After seven years in Richland, Jim can count his genuine friends on one hand. 'The question is, should I pull the plug on this thing now or wait for someone else to do it for me?' The crisis is urgent: in a few months the regulators will inspect his company, and he knows they will find him in violation because he has not been able to pay for the latest set of safety applications. The inspection could result in a fine anywhere between US$50,000 and US$500,000. He should take comfort in the wise words of his artist friends who tell him that it's time to give up and come home. The work is done. He has accomplished what he set out to do in every way but one. He's turned the nuclear industry on its head, defied their rules, taught them a little about art and inspired artists throughout the country. Now is the time to pull the plug on the project. 'I have a dream,' he tells me. 'I want to move to a small town in North Dakota, pretend to be a deaf mute and work nights washing up at a Denny's Coffee Shop. I'd spend the rest of my days watching TV reruns of *I Love Lucy* with Eddie the cat in my lap.'

There is of course one hitch, his 'tar baby': the twelve brand-new, stainless steel, radioactive, fuel rods neatly stacked in boxes, inside the security fence on the Hanford site.

3. *Wired*

Gary Lekvold was doubtful about the psychiatric examination, but he went along with the doctor in any case. Dr Montgomery was a nice guy, according to Lekvold, and he seemed pretty honest. 'I have a few questions for you, Gary,' the psychiatrist began, 'and by the end of today I intend to have some answers. When we're finished, I'll let you read my report – it will save you time requesting it through the Freedom of Information Act.' Lekvold shrugged. Why not? He had nothing to hide apart from the odd quirk and foibles – same as everyone else. He wasn't *crazy*.

The questions were straightforward. Did he currently have an addiction to alcohol? Did he have a drug dependency? Was he a pathological liar? Did he have a psychotic fascination with destruction? Lekvold told the doctor everything or, at any rate, he told him about the drink problem he'd had when he was young, and that he had smoked marijuana in the past. He also recounted a dream he'd had in the third grade at school, of riding his bicycle in the sky. He didn't lie, he hates liars. In fact, if he's pathological about anything, it's about lying. Because it was lies that had got him into a mess in the first place.

There is a security fence around the perimeter of Camp David, the official retreat in Maryland of the President of the United States. It is known in the security business as the Lekvold fence. Gary Lekvold designed and developed and tested the fence – and now it protects the most powerful man in the world. Needless to say, Lekvold is proud of that fence. He's proud of a lot of things: working as a security expert for government agencies and private companies up and down the country, working for the Defence Department at NORAD, the long-range nuclear missile base in Colorado, working on the security of the Nuclear Weapons Programme for the Atomic Energy Commission and later for the Department of Energy. His c.v. boasts extensive experience in electronics, optics, mechanical engineering, high explosives and missile-range instrumentation. He has been involved in some of the government's highly classified 'black' projects, including the development and testing of the Stealth Bomber. He helped plan security for the 1984 Olympic Games in Los Angeles and has worked on security measures at a number of state prisons. Gary Lekvold is a security whiz-kid.

'I must admit,' said Dr Montgomery finally, 'I was a bit concerned about you coming here, and so was my receptionist. We'd been led to believe that you were a raving lunatic.' Prior to the meeting, the Department of Energy had sent the psychiatrist volumes of information about Lekvold. The dossiers had painted a pretty ugly picture. So when Lekvold turned up in his dark business suit carrying a briefcase, Dr Montgomery was understandably taken aback. On the day that Lekvold was suspended from duty at Hanford in 1989, the Department of Energy distributed a character profile of him to the Washington State Police and every local police station in the area. It warned the law enforcement officers, 'This individual is extremely knowledgeable of the Hanford operations and is also extremely intelligent . . . There is reason to believe that his reaction to this suspension may be unpredictable. He claims to be a weapons expert and states he has a large cache of weapons and ammunition at his residence. Date of Birth: 12th October 1942. Height: 1.75 m. Weight: 70 kg. Hair: Brown. Eyes: Blue. Vehicle: 1987 Chevrolet Camaro (silver).'

I had no idea that Lekvold was a well-armed and dangerous individual when I arrived at his cedar-panelled house in Pasco, near Richland. He was dressed in pale-blue jeans, immaculate white sneakers and a loose short-sleeved shirt that had been left untucked. His fine hair was swept forward over his cranium in an attempt to disguise the fact that it was thinning. As we talked in the sitting-room, he sat casually with one foot propped on his knee and his arm stretched out across the back of the couch. He smoked filterless Lucky Strike cigarettes, a habit he had given up, but since the trial he has smoked a pack a day. Around the walls flashed a small red light in rapid sequence. The light flitted, like a firefly, across the mantelpiece, around the windows and then disappeared up the stairs, only to reappear a second later. A high-level security sensor, it gave the unnerving impression that the house contained something extremely valuable or dangerous – or both. It was not immediately obvious, however, what such a thing might be: the room contained only a large television set, a collection of beer cans and baseball caps, a coffee-table and the couch. His most prized possession in the house is an old leather trunk that his father brought with him from Norway when he arrived penniless in New York.

Lekvold lives by himself and owns up to being lonely. It is clear that he doesn't spend a lot of time entertaining. His kitchen and sitting-room are both untouched. In the bathroom, next to the lavatory, half a dozen magazines are displayed neatly, like a deck of cards, with only the titles visible. (The magazine on top is *Playboy*.) Most of his friends refuse to speak to him these days, no doubt because they fear, or have been warned, that hanging out with Lekvold can be seriously career-limiting. In spite of the loneliness, Gary Lekvold enjoys the climate of south-east Washington. He likes the local people, he says, or at any rate, the few that don't work at Hanford. But what keeps him in the area is his house. 'I've lived out of a suitcase for most of my life, seen too many airports. I needed to settle down, and I knew the minute I saw this house that it was exactly what I wanted.'

After we had talked for an hour or so, he unlocked a door in his kitchen and swung it open. He asked me, 'Would you like to see where I spend most of my time?' Through the door, a few steps down, was a basement so spacious that it was hard to believe it belonged to the same house. Hanging on the walls in neat rows were thirty types of hammer, fifty different pairs of pliers, miscellaneous saws and every kind of cutting instrument imaginable. I couldn't help being reminded of a scene at the end of the film *Silence of the Lambs*, when the serial killer, having been confronted by a suspicious Agent Starling, disappears through a door in his kitchen. The Special Agent chases after him, and they descend into a vast labyrinth of grimy corridors and rooms full of unspeakable horrors. Lekvold had converted his garage into a workshop. On the workbench that ran the length of three walls were bolts, wires, safety masks and sophisticated electrical equipment. The floor was littered with different types of sheet metals, bits of rubbish and more odds and ends rescued from skips. Five large metal-working machines stood in the middle of the room, an area that had been set aside for cutting, welding, soldering, polishing and drilling.

'This is where I like to spend the day,' said Lekvold, showing me around. 'I make everything from my own security devices to these –' he picked up a dagger with a 20-centimetre razor-sharp blade. Such knives are popular with hunters, he explained. The blade is stronger than anything you would find on mass-produced hunting knives. That is because of the type of steel used and the design – the Lekvold design.

'Do you make guns?' I asked.

'That's an odd question,' he said, with a sheepish grin. 'Why do you think I would make guns?'

'Just a hunch.'

'Well, you'd better have a look at the rest of the house then,' he said, as I followed him back to the kitchen.

'People occasionally ask me how many bedrooms my house has. I always say that it's a one-bedroom house with a gun room and a computer room,' he laughed as we climbed the stairs. He showed me into the gun room, once the spare bedroom. The bed had gone; in its place were more metal-working tools and a large workbench. Small boxes are stacked on top of one another, containing different calibres of cartridges and shells – snub-nosed, fine point, hollow point. There are bullets for hunting rifles, semi-automatics, handguns, shotguns and containers of gunpowder. Like most gun enthusiasts in America, Lekvold 'rolls his own', that is to say, makes his own bullets to save money. One of his specialities is designing his own bullets for target practice. He spends his weekends with members of the American Gun Club, refining his technique for long-distance target practice using a handgun. The handguns are custom built to suit each marksman. They practise by firing at cut-outs of chickens and goats placed at anything between 50 and 250 metres. Lekvold shows me the target he is most proud of. The fact that the holes are so close together at that range is a remarkable achievement only possible, he explains, through the perfect harmony of the gun, the bullet and the body. It is all down to the *craigmoor* firing stance, and he twists himself on the floor in what looks like a Yoga lotus position. 'But where are the guns?' I ask. He shows me several small polaroids pinned on the wall; there are over 500 guns of every description laid out in rows. 'These are old photographs. I've got more than that now.' But when I asked where he kept them, he grinned widely, 'Now that is a stupid question,' and walked off into the computer room.

Lekvold admits to spending too much time playing computer games. His favourite, *DOOM II*, involves killing alien mercenaries in particularly bloodthirsty ways. Until the trial, the computer room belonged to Lekvold's daughter, Kimberley, who returned to live with her father after his ex-wife died in a fishing accident. Lekvold had a particularly turbulent eleven-year relationship with Kimberley's mother, which ended in an acrimonious divorce. Besides, he had not seen his daughter until after the fishing accident. 'It was very difficult for both of us because we had spent so much time apart. That woman poisoned Kimberley for so long with lies about who I was, that it took a long time getting to know each other again.' The day that Kimberley arrived at Pasco airport, Lekvold greeted her with a huge bunch of pink roses. His drinking partners at the Red Robbin in Richland put a banner across the front of the bar saying, 'WELCOME HOME, KIMBERLEY'. 'Perhaps I went a little over the top,' he says, looking around the room littered with papers. 'I made this room into the perfect bedroom for a teenage girl and bought her anything she wanted. I was just so excited to see her again.'

Gary Lekvold first visited Hanford as a security consultant in the early 1980s. At the time, the Department of Energy was installing an Israeli-designed intruder system, and he was the expert advising on the project. He was shocked by the lack of protection at Hanford. So, when he was offered a full-time position working as a supervisor at the Security Applications Center, the security think-tank, he felt that there were a number of ways in which he could contribute. He accepted the offer and, in 1985, moved permanently from Albuquerque. A few months later, he was asked to carry out a vulnerability assessment of the 'K' basins. It was a high-profile assignment. The DoE was concerned about the security of the irradiated fuel assemblies stored in water in the basins. It had already completed a preliminary study, but the new report was intended to go into more depth and to make a final assessment. Lekvold concentrated on the physical security of the basins, while his co-author examined the basins' contents. The two of them had worked together on a variety of projects in the past and, although not a perfect working relationship, according to Lekvold, it was adequate.

A few weeks later, however, Lekvold's colleague withdrew his co-operation and refused to let him see any of the findings. The section of the report that was kept under wraps referred to the 'self-protecting' nature of the fuel rods in the basins. In other words, were they so 'hot' that it would be impossible to remove them without risking certain death? At the time, the Department of Energy was considering a proposal to reduce the level of radiation that was officially defined as 'self-protecting'. Or, as Lekvold believed, were the 'K' basins insecure because 98 per cent of the fuel rods were not 'hot' enough to deter potential thieves. By the time the report was complete, Lekvold had still not read his colleague's assessment. He complained to his manager, demanding to see it, but his request was turned down. In any case, his unilateral action led to a fierce argument with his co-author in the office one day. 'He was really mad,' Lekvold recalls. 'He was waving his arms about, calling me everything under the sun. I was convinced he was going to hit me.'

Not long afterwards, a group of security managers were invited to report on the findings at the Department of Energy headquarters in Washington DC. It concluded that security at the 'K' basins was adequate, in light of the fact that several million dollars had already been spent making it so. When Lekvold eventually got to read the report, as it was handed to the Department of Energy, he was outraged by his collaborator. 'He had written things that were false. It was like black and white, hot and cold. I immediately insisted that my name be removed as co-author of the report.'

It sounded a bit far-fetched. Why would anyone want to steal a spent fuel rod? I wondered if he was being somewhat over cautious. 'Listen,' he replied, 'there were about 360 active terrorist organizations in the world at the time. Acts of terrorism happen every day, it's just the big ones that get reported. I calculated that it would be very easy for five dedicated people with a budget of US$10,000 to steal, say, ten fuel rods from Hanford, and there is a good chance that not a single person would ever know about it. Think of the thousands of Arab terrorists who are willing to die for their cause. It would be a snap.' He thought for a second. 'You give *me* ten irradiated fuel assemblies and watch what I do.' He pondered again. 'With a case of dynamite you could threaten to spread that stuff around New York City. You'd cause mass panic.'

As soon as the vulnerability assessment was over, working relations in Lekvold's office deteriorated. After a few weeks of being given the silent treatment by his colleagues, Lekvold requested, and was granted, a transfer to a Westinghouse security outfit working next door. After one week into the new job, his manager gave him a copy of his personnel file to read. 'My mouth dropped open when I read it. It attacked my credibility, my competency, my character, honesty and my ability to work with others. It was a total hatchet job, pure garbage.' With the knowledge that his career at Westinghouse was about to be cut short, Lekvold sought the support of a security director at Hanford. He provided him with sixty pages of documents that he had collected as evidence relating to his allegations. 'He appeared to be very concerned and concluded that, if I was willing, he would assign a first-rate investigator to the case.' Lekvold admits that alarm bells should have rung the moment he discovered the investigator was an ex-employee of the Rockwell corporation. Most of the managers he was making allegations against were also ex-Rockwell. The investigator assigned to the case by the security director met with Lekvold on several occasions. Together they pored over hundreds of papers, going into every allegation in the minutest detail. Lekvold felt pleased with the way things were going until it began to dawn on him that no action was being taken and there was little likelihood that any would be. A fortnight after meeting Lekvold the investigator went to work for one of the managers who had been targeted in the investigation. 'I have absolutely zero respect for that so-called investigator. I would tell you how I feel in more detail, but I'm afraid I'd be libelled,' he says. 'He is a very dangerous person. He's been instrumental in the other whistleblower cases. I will say, though, he is good, he did a good job of destroying me. Oh, yes, he's smooth, he's clever and now I hear he's rising quickly through the ranks at Westinghouse.'

Over the next few months Lekvold met with other Westinghouse officials. He made it perfectly clear that he had no wish to go outside the company with his allegations. The director of security flew in from Pittsburgh and met him on five different occasions. He also saw the head of personnel. At his final meeting, according to Lekvold, the security director shouted obscenities at him and implied that he had mental problems. He also told him to see the Employee Concerns officer, who immediately referred him to the company psychiatrist. In desperation, towards the end of 1988, Lekvold delivered Westinghouse an ultimatum. He demanded that if they did not meet his requests to improve security at Hanford, he would be forced to go to the Department of Energy. That week a top-level meeting at the Richland headquarters discussed the Lekvold case. The ultimatum stipulated that a decision had to be made by the end of the working week. At 4.58 p.m. on that Friday, a brown manila envelope was left at the Westinghouse reception for Gary Lekvold. The letter inside informed him that his demands could not be met.

Lekvold was forced to take his complaint to the Department of Energy. Yet another investigator was assigned to the case. In the course of a DoE meeting, Lekvold was notified by Westinghouse that he was immediately suspended for insubordination from any further duties. 'The specific allegation against me was failure to fill out a time-card,' says Lekvold. 'To this day, that is the only official charge against me. I ask you, does it sound normal for someone to be suspended for twenty-three months on full pay for refusing to fill out a time-card? I think not. They simply wanted me out of the way because they knew that every day I was in the office I was watching what was going on and taking too many notes and finding out too much.'

No longer having a job, Lekvold spent his days building the workshop in his garage and working with his lawyer on an appeal. One afternoon he noticed a car parked at a T-junction about a kilometre up the road. It had been sitting in the same position for several hours. Reaching for his binoculars he focused on two suited men sitting in the car and discovered, to his amazement, that they were doing exactly the same to him. Gary calculated that from where they were positioned he could sneak out the front door into his car and still be out of their field of vision. He pulled slowly out of his drive and instead of turning left at the end of his street and heading towards Richland, he turned right and did a wide circle bringing him right up behind the car sitting at the T-junction. As he approached, he could see the two men seated in the car with their binoculars still trained on his house. He noticed that the car had government licence plates. He then pulled up next to the car and as he came alongside, he turned to the two men inside, grinned, leaned out of the window, photographed them and then sped off leaving a cloud of dust behind him. In his rear-view mirror he could make out the pandemonium in the car. With the realization that they'd been caught red-handed, the two men scrambled to start up the engine and disappeared in the opposite direction.

Lekvold dismisses the incident as just another childish game played by Westinghouse, but he admits it is very easy for a whistleblower to develop paranoia. He calls it BMS, Bogey Man Syndrome. But he also knows that these people can play nasty games. He has heard stories of people being followed, threatened and having their property broken into. One of their favourite ploys, he told me, is to place a piece of paper on the dashboard of your car with the words, 'YOU ARE BEING WATCHED', written on it. He found the same message among some papers at his office. For several weeks he was convinced that his phone was being tapped by security officers at Hanford. On several occasions when he picked up the phone he heard a strange single bleep. (It was only later that he discovered that his telephone number had been printed mistakenly as the fax number of a hotel in Pasco.) 'A strange thing did happen to me,' he says. 'I was calling my lawyer one day, when there was the odd background noise and sort of echoing on the line. To be on the safe side, I said I'd call him back and put the phone down. I drove into Pasco to a phone booth in the middle of a parking lot at the mall and redialled. As I was talking, a plain unmarked car came down the street and circled the phone booth no more than two metres away. The man inside stared at me as he circled two or three times and then drove off. The thing was that I recognized the man, he was a Hanford patrol officer. How did he know that I was there? You can only come to one conclusion.'

A month after his suspension, Lekvold was notified that his 'Q' clearance had been randomly selected for a routine check. 'I've had my "Q" clearance since 1966,' he told me. 'Before that I had a secret security clearance at the Defence Department. In all that time it has never been rechecked, and they wanted me to believe that my number "just happened" to have been randomly selected.' The Department of Energy holds internal administrative review hearings for cases of this type. It is not a court hearing and there is no due process involved. The judge at a hearing is normally a retired intelligence agency official, who is employed by the Department of Energy on a case-by-case basis. This conflict of interests undermines the judge's impartiality. A typical review hearing lasts three hours and involves about three character witnesses and a hearing council, which presents all evidence to the person under review. Lekvold's review hearing was held at the federal buildings in Richland and in Albuquerque, and took four weeks; five days a week, eight hours a day. It is estimated to have cost the Department of Energy one million dollars. His hearing council, under the guidance of Lorette Enochs, the Department of Energy's chief lawyer, called forty-six people as witnesses.

'Enochs spoke to everyone I'd ever known. She trawled through my entire life in the minutest detail,' says Lekvold. 'She made a big issue of the fact that I had broken a window in a disused gas station when I was seven. She spoke to the people at my local 7–Eleven store to find out how many lotto tickets I bought each week. She quizzed former girlfriends about my sexual preferences and performance. She talked to every cocktail waitress and bar owner in town about my drinking habits.'

Tom Carpenter, Lekvold's lawyer, says, 'Lorette Enochs was like a pit bull unleashed on fresh meat. She acted like a zealot out to get him. The whole hearing was straight out of Kafka. Gary was accused of all these things and had no idea what they were going to come up with next. Most of it was pure and utter lies, and they got all these people to perjure themselves.'

'One waitress testified against me,' says Lekvold. 'But everybody knew that we didn't like each other, it was no big deal. She told the hearing council that I would go into the Red Robbin after work, hammer down six whiskeys in rapid succession and then start making a fool of myself, falling off my stool and stuff like that. She said that she had to sober me up by serving me cold tea instead of whiskey and that I was so drunk I couldn't tell the difference. Well, the joke of it was that I don't drink whiskey, never have. I don't like it. I always drink beer, as everybody knows.'

In the course of the hearing Lorette Enochs turned up unannounced in his backyard one day. She was escorted by the head of security. Enochs explained to Lekvold that he had to see a company psychiatrist. He first began to correct her, 'You mean a psychologist.'

'No,' she replied, 'I mean a psychiatrist.' She gave him a choice of three doctors. He chose Dr Montgomery.

Dr Montgomery's report on Lekvold's mental state gave Lekvold a clean bill of mental health. But the Department of Energy, unhappy with its verdict, ordered a second meeting with Dr Montgomery. This time they supplied the doctor with more information about Lekvold. For five hours Montgomery put Lekvold through computer tests, blood tests and psychological tests. At the end of the session he wrote a more detailed report confirming what he had said in the earlier one. Lekvold's mental state was 'normal'. Predictably, Lorette Enochs was angered by the psychiatric report: she had based her case on proving that Lekvold had an obsessive compulsive personality. Before the hearing, she had said to Tom Carpenter, 'You wait till we get Montgomery on the stand, we'll blow your client out of the water.' Dr Montgomery was eventually called as a witness for the defence. He stood in the witness box for two hours while Enochs questioned him. Lekvold remembers counting twenty-one questions in a row that ended, 'Now, doctor, would that not prove an obsessive compulsive disorder?'

'No, ma'am, it would not,' the psychiatrist repeated. The questioning ended because the judge finally lost his temper with Enochs, saying, 'Now wait a minute, let's just see who's being obsessive compulsive in this courtroom?'

'They tried to pin everything they could on me, from being unpatriotic to shooting a neighbour's pet, and at the end of the four weeks it all boiled down to one thing, that I had smoked some pot in the 1960s and 1970s.' The last time Lekvold says he smoked marijuana was at a Department of Energy Christmas Party, with forty other people who had 'Q' clearances. 'I've smoked pot with DoE security officers. I've smoked it with the vice-president of a DoE contracted company, I've smoked it with security officers who carry guns. Everyone has smoked it at one time or another. The President of the United States has smoked it, for Chrissake! But when you're filling out your "Q" clearance forms, you deny that you've ever taken drugs. Of course you do. You'd be crazy not to.'

On 13 November 1990 Gary Lekvold was told that his 'Q' clearance card was being withdrawn. It was a subtle move. A security expert without a 'Q' clearance is unheard of. His subsequent lack of status gave Westinghouse the perfect excuse to dismiss Lekvold. He lost everything: his job, his freedom to work in the security field, his reputation, his friends and his money. In a fit of anger he swore publicly that he was determined to wreak revenge, and that he would not rest until he had destroyed the careers of his enemies. At the beginning of 1995, after several years' unemployment, Lekvold got a job as a gamma-ray logger at the liquid-waste tanks on the Hanford site. Two other Hanford whistleblowers work for the same contractors. Lekvold had a struggle getting the required 'L' clearance for the job – the lowest level of access to the site. Every week he sees the people who successfully ruined his career. Nothing is said. Lekvold met a security guard at a bar one night. They got chatting and he told Lekvold that thanks to him there had been a lot of changes for the better on the site.

There is another consequence of the hearing that Lekvold would rather not talk about. Lorette Enochs approached his daughter, Kimberley, and persuaded her to testify. Enochs used her testimony to attempt to prove that Lekvold had an obsessively dependent relationship with his wife, akin to a drug or alcohol dependency. Since that day Gary has neither seen nor spoken to his daughter, and he doubts that he ever will again.

Lorette Enochs no longer works for the Department of Energy. Gary Lekvold is currently suing Westinghouse for US$10 million.

PART TWO

4. *Rocky Flats*

In the early hours of 6 June 1989 seventy agents from the FBI and the Environmental Protection Agency raided Rocky Flats nuclear weapons plant outside Denver, Colorado. The investigation of the plant had been so secret that the raid took even the state governor by surprise. Over the next ten days, as the FBI agents loaded boxes of documents into the trunks, Americans watched transfixed by the extraordinary event unfolding on their television sets. It was the first time in the history of the United States' involvement with nuclear weapons that the government had criminally investigated and raided its own backyard.

In an affidavit published after the raid it was asserted that the plant was under investigation because there was cause to believe that Rockwell and the Energy Department officials knew that Rocky Flats did not comply with environmental laws and regulations and that they concealed Rocky Flats' serious contamination. The FBI confirmed that it had conducted night-time observation flights over the site. Using infrared cameras, the pilots had identified that an incinerator, which had been shut down for safety reasons, was being used at night – and in secret – to burn nuclear waste. Other infrared photographs showed evidence of heat rising off one of the evaporation ponds, proving that the pond was being used as a disposal site for liquid waste less than ten days after the Environmental Protection Agency had outlawed the practice. Two years earlier the Department of Energy (DoE) had awarded the Rockwell Corporation, its site contractor, US$8.6 million, as a bonus for excellent management. The award was made in spite of the knowledge that Rockwell was breaking environmental laws. Upon publication of the affidavit, the Department of Energy refused to comment, except to say that public health and safety had at no point been jeopardized. Nevertheless, the city of Broomfield, 30 kilometres away, made an attempt to divert the

Great Western Reservoir, its main water-supply, from Walnut Creek, which had been contaminated, according to the document. The city was forced to suspend its action until a permit was granted by the Environmental Protection Agency.

Two and a half years later, after the district court had heard testimony from 110 witnesses and examined 760 boxes of documents, the chief federal prosecutor, Michael Norton, announced that the Rockwell Corporation would plead guilty to ten charges under the Resource Conservation and Recovery Acts and the Clean Waters Act. Norton agreed with the corporation that, if the Department of Justice waived violations of the Clean Air Act, Rockwell would pay a settlement of US$18.5 million, the largest fine ever collected by the Federal Government for a waste-disposal violation. It was also the first time that a contractor had paid the fine with its own money. (Normal procedure in the event of a fine is that the DoE pays.) At first the settlement appeared to be a significant victory for public health, a clear warning to American industry that it was perilous to disregard environmental laws. But for those who had followed the case there were many unanswered questions. Why was the Rockwell Corporation's fine less than the amount it received in bonuses from the Department of Energy over the fourteen years of their contract, a figure that amounted to only 3 per cent of the corporation's profits for 1991? Why had there not been a single indictment of any Rockwell employee at the plant or of any Department of Energy supervisor?

Denver is the rich, white-collar capital of Colorado. For decades this flat city has been sprawling outwards, swimming pool by swimming pool, garage by garage. The city's population of 2 million people continues to grow each year as thousands of Californians head east in search of clean air, a crime-free environment and the Rocky Mountains. House prices in the suburbs start at US$175,000. During the Carter Administration, anyone buying homes in Denver was formally notified, in advance of signing on the dotted line, that one of the country's largest nuclear sites lay within 25 km of the finely sprinkled lawns. As soon as Ronald Reagan got into power, it was decided that such a practice undermined the feelgood factor – and so it stopped.

Today, most new arrivals are unaware of their potentially explosive neighbour.

In 1951 the 28 sq. km of flat prairie, known as Rocky Flats, north-west of Denver was chosen by the Atomic Energy Commission (AEC) for more or less the same reasons people move there today. The land fitted government specifications of a dry, moderate climate, with close proximity to a population of at least 25,000. But according to local attorney Howard Holme, who was involved in a landowner's suit over the site, little was done to find out if Rocky Flats was environmentally suitable as a nuclear plant. Several members of the AEC suggested that it was too close to Denver, but they were overridden, and before state officials could have a say, the decision to start building had been made. 'THERE'S GOOD NEWS TODAY, US TO BUILD $45 MILLION "A" PLANT', read the headline in the *Denver Post* on 23 March 1951, and for the next forty years Rocky Flats manufactured plutonium triggers (or 'pits') for America's nuclear stockpile.

At Rocky Flats there is an 'infinity' room that is so contaminated that no equipment is able to read the radiation levels. (Standard equipment reads alpha radiation at a million counts per minute.) In all, Rocky Flats has sixteen 'infinity' rooms contaminated with plutonium that are estimated to be 25,000 times the level of background radiation. The doors to each room are permanently locked and sealed not only to prevent its contents escaping, but also to stop people going into the room by mistake. 'The gaps in the door were sealed with duct tapes,' observed Bob Alvarez, of the Department of Energy, when he told me about one such room. 'Through the small perspex window in the door I could see a large, messy room, the size of a small warehouse. There were some papers lying scattered on the floor, a typewriter was sitting on a desk and a metal-framed swivel chair had been tipped over. It all looked pretty innocent, but taped to the outside of the door was a small note which reads, "Pu238 10 x 6 MPC", which means that the room contained approximately a million times the maximum permissible concentration of plutonium-238. It was a pretty chilling thought that what separated me from that was a sheet of perspex.' A Department of Energy report, published in 1994, painted Rocky Flats as an ageing factory in terminal decline. The plant, it reported, has 12.8 tonnes of unstable and potentially explosive plutonium in over 27,000 canisters, plastic bags, bottles, metal cans and drums. Tanks and cracked pipes leak plutonium-nitrate solutions into operating areas. The radioactive glove boxes, used for handling the plutonium, corrode the windows, gaskets and rubber gloves that insulate the workers. Plutonium-nitrate and nitric-acid spills have contaminated floors and equipment. Over 8,000 drums of waste and plutonium scrap are stored in the buildings. There are unknown quantities of plutonium in the site's ventilation ducts. Many of the building structures are weak and could not withstand a large earthquake or a strong wind. At the time of the report there were over 2,000 outstanding requests for the installation of vital safety systems, some dating as far back as 1988. 'The way the Department of Energy turned off the nuclear weapons complex at the end of the Cold War is like turning your car off when you are going down a road at 100 km an hour. You suddenly lose total control over the beast,' says Arjun Makhijani, of the Institute for Energy and Environmental Research. The Department of Energy estimates the 'Cold War mortgage' for Rocky Flats at US$22 billion, over a period of sixty-five years.

Jim Stone has had a long association with Rocky Flats. He worked for the Austin Company in 1952 as a chemical engineer, on the original designs for the plant. In 1962 he returned there to work as a sub-contractor, at the site's Critical Mass Laboratory. In his last job, beginning in 1981, Jim Stone worked on the remodelling of the ageing plant, for the Rockwell Corporation, as part of President Reagan's Administration's 'rearming America' initiative. 'I was extremely proud to go and work for Rockwell,' Stone explains. 'They had built the Challenger and were highly respected in the industry. It was a wonderful opportunity to do something for God and the world.' After six months at the plant Stone was promoted as a principal engineer in the Utility Division, with responsibilities which included the writing of progress assessment reports. Stone, however, began to realize that his reports were being ignored. He came to view the Rockwell Corporation in a different light. Stone raised his complaints with a Department of Energy site officer. But instead of notifying the DoE headquarters in Washington, the officer went straight to the Rockwell bosses. Jim Stone was ordered to a meeting at the head office. Shortly after, a number of redundancies were made and Stone was among the first to go. 'I was called in one morning to my manager's office and told that I had one hour to pack my things and leave. They were real gleeful about getting rid of me.' Stone returned to his desk and, under the gaze of a guard assigned to escort him off the premises, he packed his belongings into boxes and left Rocky Flats. But the guard failed to notice that Stone walked out of the main gates carrying his reports and papers.

In anger, he turned to a new law designed to help government departments keep a tighter control on their contractors. The Federal False Claims Act allows a person to sue a contractor on the Government's behalf and share in the settlement, if successful. Instead of taking his allegations to the Department of Energy for a second time, Jim Stone went to the FBI. He met with the environmental investigator, Special Agent Jon Lipsky. 'I laid all my files and reports out on Lipsky's desk. He couldn't believe his eyes. He had never seen so much information and so much detail from a whistleblower before.' Over the next few days Jim Stone told Lipsky everything he knew. After that, however, he was never contacted by the FBI again.

A grand jury was appointed two months after the raid on Rocky Flats. The twenty-three members of Special Grand Jury 89–2 were picked at random from the Colorado voting and driving-licence registers. Wes McKinley, the jury's foreman, was a cowboy who organizes trails for 'city slickers' in the south-east corner of the state. The other members of the jury included a swimming instructor, a bus driver, a secretary, a bartender, a hairdresser and a retired sheriff. For the next two and a half years this group of people met in Denver, one week out of every month, in what was to become the longest environmental crime investigation in American history. The details of what went on in the courtroom have never been made public, and to this day each member of the grand jury is under threat of imprisonment if he or she talks about the evidence.

By the autumn of 1991 most of the 110 witnesses had been heard and the jury began to sift the evidence. Ken Fimberg, a US attorney and Norton's assistant, told the jurors that in his opinion there was enough proof to indict ten of the Rockwell employees. But indictments of DoE supervisors, he added, would not be appropriate, because the government agency was too extensively involved as a whole. As an example, he pointed to the fact that the DoE had given Rockwell permission to incinerate waste in 1988, even though it was understood that Rockwell did not have the necessary permit to do so. The jury, however, was not convinced by his argument. According to the jurors, the DoE was directly involved and its employees were just as culpable as the Rockwell employees. They were in favour of indicting three of the DoE's supervisors. Explaining the recommendation, the jurors wrote in a final report that, 'Criminal conduct should never be a part of a government employee's work. If the government employees do not obey the law, we cease to be one nation under the law.' This disagreement between the grand jury and the prosecution council came to a head in November 1991 when Michael Norton, on hearing that the jurors wished to pursue charges, announced angrily that he would not sign any indictment he had not drafted himself. Moreover, he would not be indicting any DoE or Rockwell employees. Norton strongly advised the jury not to meet again. The following month Ken Fimberg unexpectedly announced to the court that the prosecution had finished presenting its evidence to the jury, at which point all three attorneys for the prosecution got up and walked out of the courtroom, leaving the jury no idea how to proceed. Its members were so angered by Norton's refusal to sign an indictment written by them, that they thought about hiring a lawyer to investigate him for obstructing their work. It seemed patently obvious to them that Norton had come to a decision not on the basis of the evidence presented in court, but as a result of political pressures from his superiors at the Department of Justice. 'Why should the prosecution have changed their minds about indicting individual Rockwell employees, if they hadn't all of a sudden gotten orders from above?' asked one juror. Michael Norton, the jury protested, had refused to subpoena a witness they wished to question a second time. On another occasion he had directed a witness not to answer certain questions from the jury.

With the prosecution now absent, the grand jury looked to Judge Sherman Finesilver for guidance. They had a number of questions. Could an indictment be issued without the attorney's signature? If so, would it be valid? Could a government attorney be forced to sign the indictment? Judge Finesilver was not expansive in his answers. An indictment, he told the jury, had to be signed by the attorney. He referred the jury to the book of instructions handed out at the beginning of the trial. The rules stated that jurors could issue a 'presentment' without the attorney's signature.

Shortly before Christmas of 1991, the members of the jury were alarmed by a message from Judge Finesilver, in which he congratulated them for their work and then dismissed them from any further duties. In response, Wes McKinley, the jury's foreman, wrote a letter to the court. It began, 'In the month of August 1989, Judge Finesilver gave Special Grand Jury 89–2 an obligation. Special Grand Jury 89–2 was instructed to look out for the best interests of the people of Colorado and the national interest. Special Grand Jury 89–2 is very serious about fulfilling this obligation and fully intends to complete the duty it was given responsibility for.' The letter concluded by demanding that a further session be organized during the third week of January. The court acceded to their request, and in January the jury drafted three documents: an indictment, charging individuals at both Rockwell and the DoE, in the hope that another attorney would be appointed to the case; a presentment, almost identical to the indictment which they hoped Judge Finesilver would release to the public; and a report of non-criminal activities alluding to the conduct of DoE, Rockwell, state and federal regulators at the time of the original raid. On completion, the documents were deposited in a court-house vault. Under the court's rules jurors were not allowed to keep copies of what they had written.

The documents charged eight Department of Energy and Rockwell employees with criminal behaviour. These included two managers of the DoE's field office, responsible for the oversight of the Rocky Flats plant. The DoE's area manager for Rocky Flats was also charged, among other things, with making false statements in a letter he signed in 1988. The letter, drafted by a Rockwell employee, stated that evaporation ponds at the plant had not been used in the last twelve months. Other Rockwell staff charged were the director of Waste Operations, who had told a State Health Department officer that the evaporation ponds were not being used, even though he knew that they received waste regularly. The director of Plutonium Operations was also charged, as was the supervisor of Environmental Compliance Programs and three mid-level managers, including one who wrote the letter signed by the DoE area manager. The report went on to depict the Department of Energy as feckless, incompetent and at times duplicitous. 'The DoE,' the report stated, 'explicitly discouraged Rockwell from complying with environmental laws, by omitting environmental compliance from the list of criteria according to which large performance bonus fees were paid to Rockwell. Significantly, these large financial incentives could be earned most easily if Rockwell ignored environmental compliance in striving to meet weapons production goals . . . Rockwell conspired with certain DoE officials over a period of years to hide its illegal acts and the illegal acts of its employees behind the sovereign immunity of a department of the Federal Government. The DoE continues to direct and endorse this course of illegal activity in violation of applicable environmental laws and in the name of political expediency.'

On 24 March 1992, the grand jury filed into the courtroom for the last time. The atmosphere was tense. Michael Norton held a copy of the jury's report in his hand. It remains a mystery how Norton got hold of the report. In normal circumstances he would not have had access to the courtroom vault. In any event, the day's proceedings started with Norton attacking the report's inadequacies and amateurism. He told the court that there was insufficient evidence to support the report's allegations. He then provided the jury with a copy of his own indictment and asked the jurors to sign it. They read the indictment but refused to sign on the grounds that its findings were equally inadequate. It was a stalemate. In the afternoon Judge Finesilver announced that the jury was dismissed. Two days later Michael Norton held a press conference in which he stated that Rockwell had pleaded guilty to ten charges and had agreed to pay the record-breaking figure of US$18.5 million. On the day of the announcement

Rockwell's stock closed at US$23.75, up 12.5 per cent. The settlement was accepted by Judge Finesilver three months later, and a move to release the jury's reports was refused. The Rockwell Corporation had been replaced as the contractor at the Rocky Flats plant, but its bosses were unrepentant, maintaining that there had been no risk to public health.

In the following weeks a batch of newspaper articles expressed concern about the Rocky Flats settlement, but Michael Norton maintained that he had secured a record-breaking deal and that there was not enough evidence to indict individuals. 'Criticism,' he said, 'is coming from people who are fundamentally ill-informed, who write for partisan purposes and have their own agendas.' In response to the questions being asked on Capitol Hill, a house committee on science, space and technology (in Washington DC) voted to subpoena documents in order to find out the truth. Tersely worded letters were exchanged between the committee and the Justice Department, which claimed immunity to answering questions about what it referred to as its 'prosecutorial discretion'. The chairman of the committee, Howard Wolpe, a Democrat from Michigan, stated that the investigation was necessary because there was an obvious conflict of interest inside the Government with regard to how vigorously the case was pursued. A more serious conviction would not only prevent Rockwell from keeping other government contracts; it would also cast the Department of Energy in a bad light. A conviction, he went on to say, would expose the agency and company to civil suits from neighbouring towns. The Justice Department eventually decided to allow Michael Norton and the FBI agent, Jon Lipsky, to be questioned by the house committee, on the basis that they would not answer questions regarding 'internal advice, opinions or recommendations' in connection with the case. In a tantalizing session, Lipsky told the committee that he knew but he had been ordered by the Justice Department not to say why there had been no indictments. The committee, its appetite now whetted, voted to demand the intervention of President George Bush. The White House demurred, however, and for a while it looked as though the case had once again reached stalemate. Indeed Wolpe concluded the investigation by saying that it was startling that no individuals were prosecuted. He did, however, criticize the Justice Department for its belief that the individuals cited had merely adhered to a culture of ignoring environmental laws. 'The culture,' Wolpe said, 'was an institutional policy that emphasized weapons production over environment health and safety concerns.'

Meanwhile, back in Denver, Judge Finesilver had written a letter to Michael Norton expressing his belief in the maintenance of secrecy. He stressed that it was a point of law. Finesilver went on to call an investigation to discover which juror had broken the law by giving a draft of the report naming the DoE and Rockwell employees to the local weekly paper, *Westword.* On 18 November 1992, infuriated by Finesilver's suggestion, twelve members of the grand jury held a press conference on the steps of the court-house. They released copies of a letter written to President-elect Clinton asking him to appoint a special prosecutor to investigate the handling of the case. The letter was never answered.

In July 1995 Sherman Finesilver unexpectedly resigned from his life tenure as a federal judge. The appointment of his successor could shed a different light on the Rocky Flats situation; if, that is, the new federal judge decides to release the grand jury's report, which is still locked in a courtroom vault in Denver, Colorado.

5. *In Search of Wendell Chino*

> 'Apache's were free in a land that belonged to no one and yet was to be used wisely by all.' (Inscription on the wall of the Mescalero Apache Cultural Center, Mescalero, New Mexico)

I walked into the Fuel Storage Information Center in Mescalero, New Mexico.

'Yes?' said a voice. 'Can I help you?'

So I explained to the middle-aged woman behind the desk that I wanted to see the exhibition.

'Are you a reporter?' she asked suspiciously. I told her that I was writing a book, and she frowned – it wasn't quite yes or no.

'The floor's wet,' she said, because the lino had just been mopped; it was drying visibly in front of us, in the warmth of the early afternoon. 'Could you come back in an hour?' I nodded. But as I walked away a thought occurred to me, so I returned to the doorway. The woman was on the telephone. As I reappeared she immediately put the phone down.

I asked, 'You'll be open in an hour, will you?'

'I might close early this afternoon,' she said vaguely.

'Are you worried about the floor or whether I'm a reporter?'

'Uh, huh,' she replied. 'How do I know if you are a tourist or a reporter?'

'I just want to see the exhibit.'

'You can't take photographs or anything and you can't write anything down.'

I agreed to her conditions, and she told me to come back in thirty minutes. Half an hour later I returned, but the Center was closed.

There are two distinct but prosperous communities in the Tularosa Canyon between Roswell and Alamogordo, New Mexico. Both occupy the beautiful wooded pastures on the western slopes of the

Sacramento mountain range, an area known as the Land of Enchantment. This is the home of Smokey the Bear and Billy the Kid. In the long hot summers the cool air and the lush green fields come as welcome relief after the flat baked desert of southern New Mexico. As a result, the Land of Enchantment swarms with holiday-makers, and the residents have gone out of their way to encourage the influx of tourist dollars. One of the communities is Ruidoso, a mix between Las Vegas and Montana – a playground for rich tourists who enjoy a mixture of fresh air and gambling. The town luxuriates in crazy golf-courses and full-scale mock castles, but the local residents recently voted for a ban on skateboarding and bicycles, because it was feared that such behaviour might lead to hooliganism and other criminal activities. The other community is made up of 3,000 Mescalero Apache Indians, who live on a reservation to the west of Ruidoso. The tribe is famous for its motto, 'The Navajos make rugs, the Pueblos make pottery and the Mescaleros make money.' An information handout warns that 'a visitor expecting to see braves and women in buckskins and blankets' is likely to be disappointed because 'Changes have occurred in their way of life over the last hundred years.' The changes have come in the form of a ski resort, a luxury hotel and gambling centre, restaurants, a bingo hall and a golf-course, all of which – it must be said – is extremely unusual for a Native American Indian reservation. Nevertheless, it is agreed by most people on and off the reservation that these changes have been made possible by the vision, business acumen and sheer determination of one man, the President of the Mescalero Apache Tribe, Wendell Chino.

Few people on the reservation know anything about Wendell Chino. It is whispered that he is seventy-one years old. Nobody knows how much he earns, but everybody knows that he lives in the largest and most expensive house on the reservation and flies to meetings in a Lear Jet. He is occasionally referred to as 'Swindle' Chino, but only as a joke. His wife Rita runs the exotic Inn of the Mountain Gods, which advertises itself as New Mexico's most distinguished resort. Their son is the reservation's chief of police. In December 1964, critics argue that Chino amended the tribe's constitution to give himself, the newly appointed president, total control and veto over the other nine members of the tribal council. With such changes he seized the destiny of the reservation and placed it in his own hands. On a number of occasions he has been seen to walk into Casino Apache, the tribe's newest venture, and scoop thousands of dollars on the slot machines. Various people have 'gone missing' from the reservation – or so the story goes – as a result of displeasing Mr Big. 'Wendell Chino died of a heart attack in 1979,' Joseph Geronimo told me. 'What possesses his body now is the devil.'

Chino has enjoyed the respect of his tribe for nearly three decades, but suddenly his future as tribal leader is uncertain. A sense of unease and distrust of the leadership has come bubbling to the surface in recent years. It focuses on his latest venture, which is showcased by the Fuel Storage Information Center. The idea is to use a remote part of the reservation known as Three Rivers to store nuclear waste. Many people on the reservation feel that this time Wendell Chino has gone too far. In 1991, the US Federal Government, in search of a temporary solution for the storage of spent nuclear fuel, offered serious financial benefits to any state or Indian reservation (Native Americans have their own constitution and sovereign rights) willing to discuss acting as host to the waste. The final solution, a Monitored Retrievable Storage (MRS) facility, was a complex of warehouses as big as a shopping mall and able to store 10,000 tonnes of high-level radioactive waste for up to forty years. Wendell Chino was the first to apply for the MRS scheme. In late 1993, however, the government withdrew the offer after it was rejected by the other parties involved. Three years on, Chino is negotiating a US$250-million contract for radioactive waste with thirty-three private power companies.

'Wendell Chino was a minister at the Reform Church when I was growing up on the reservation,' Rufina Laws told me. 'I wanted to be a missionary, so I looked up to him. He was my hero.' Rufina was born on the reservation and grew up in what she describes as abject poverty. During the 1940s she lived with her mother in a tepee under a pine tree. (It is now the site of the bingo hall.) There was no electricity or running water. She recalls her mother making blouses for her out of flour sacks and telling her bedtime stories in the dark because they couldn't afford kerosene for the lamps. In the early 1950s 'the man I call my father', an Englishman called William Arthur Cansler, married Rufina's mother and lived with them. It was the reservation's first interracial marriage. Cansler planted crops and grew vegetables – and slowly life began to improve for the family. They moved to the more prosperous White Tails area of the reservation, where Cansler opened a candy store and a petrol station, while Rufina's mother cooked for the day-care centre. It was in White Tails that Rufina first met the diminutive Wendell Chino. She listened to his sermons and remembers his being the only minister on the reservation who could hold her attention. 'He would keep me awake by emphasizing certain words. He was a very good orator.' Wendell Chino, then in his early thirties, was already a prominent member of the community. He was intelligent, a good communicator and he came from a respected family. It was clear that his reasons for being a minister were as political as they were spiritual. He understood that the only way to improve life on the reservation was to do business with the neighbouring community. In order to learn the white man's ways, Chino turned for guidance to William Cansler, the local paleface. Rufina remembers Chino coming to the house at the end of the day and talking business with Cansler for hours. In the mid-1950s Chino gave up his ministry and went into politics full time. 'He was the only man on the reservation who openly talked of the injustices that we had suffered at the hands of the white people,' according to Rufina Laws, 'the only person who spoke about us bettering ourselves by using what we had to our own advantage.' Chino quickly began to prove himself as a leader, as a dominant force on the reservation. He managed to win the respect not only of his contemporaries but also of the reservation's elders.

In 1957 Rufina, then aged twelve, went on a day-trip to Alamogordo with her mother and a friend. In spite of her mother's protests Rufina insisted that they eat at a local restaurant, and so for nearly an hour they sat in the restaurant, totally ignored by the staff who were busy serving the white families around them. 'Have you had enough yet?' her mother asked. The experience came as a profound shock to the young woman. 'I was so angry. It was the first time that I realized my "place" in society. I remember wishing that I was white.' The event marked the beginning of Rufina's interest in the Civil Rights Movement, spearheaded at the time by the Reverend Martin Luther King. 'I began to see Wendell Chino and Martin Luther King in the same light,' she recalls. 'I thought Chino could do for the American Indian Movement what Martin Luther King was already doing for the black community. They were both ministers. Both were politically motivated and were able to communicate with large groups of people. I really believed that Chino could make our constitution more democratic, and I remember wanting to stay on the reservation in order to work with him towards that end.'

The debate about the nuclear dump has drawn attention not only to Chino but also to its opponents. The tribal council has refused to talk to the press for several months now. It is assumed that any publicity is bad for business, not least because the licensing of a site is considerably more difficult to achieve if a community is divided. I had been told on a number of occasions that Chino was extremely rude and abrupt with members of the press. I was assured that I would not be able to interview Chino nor any other tribal council member. Even so, I faxed the tribal council offices a request to interview the President – and then I waited. I heard nothing for several weeks. In the end I rang the office, hoping to pester a secretary or clerk. By chance the telephone was answered by Chino himself.

'Yes?' His tone was aggressive.

I explained who I was. I told him about the fax and my wanting to interview him.

There was a pause, and then he shouted angrily, 'I am not in the business of spending my time making other people money,' and slammed the phone down.

The first business enterprise ventured into by the Mescalero Apaches under Chino's leadership opened in 1956. It was the Apache Summit restaurant on Route 70, the main link between Roswell and Alamogordo, which cuts straight through the centre of the reservation. The restaurant, specializing in Apache-style home cooking, had a big draw: the lucky diners could be photographed with Robert Geronimo, the last living son of the legendary Apache leader and warrior. Inevitably, the venture was a success, and in 1962 Chino followed it with a more ambitious project. On the south side of the Sacramento

Mountains, an area partly covered by the Lincoln National Forest, was a modest ski-run owned by an oil millionaire who had invested in the area. Having raised US$2 million, Wendell Chino bought the ski-run and turned it into Ski Apache. The resort, now valued at US$45 million, is today one of the most popular in the country and is visited by over 300,000 people a year. The following two ventures were a lumber business, with a current annual turnover of US$4 million, and a ranch with 6,500 heads of beef cattle. Each business proudly declares itself 'A Mescalero Apache Enterprise' or 'Owned and Operated by the Mescalero Apache Tribe'. Next on the business portfolio was the Inn of the Mountain Gods, a luxury hotel worth US$20 million today, on the shore of Lake Mescalero. Nestling in the foothills of the Sierra Blanca, a 3,600-metre holy mountain to the Apache Indians, the resort is Wendell Chino's crowning glory. It boasts swimming pools, tennis courts, a cocktail bar, a casino, a shopping centre, hunting, boating and fishing, not to mention the Wendell Chino East Convention Hall and the Wendell Chino Ballroom. By the time Rufina Laws came back from university to teach on the reservation, rumours were circulating that Wendell Chino was worth over US$2 million. 'I was confused,' remembers Rufina. 'How could the man possibly be worth that much money?'

Twenty years on Rufina Laws was visiting her mother on the reservation. Rufina, by then a resident of Arizona, was putting her life back together after the death of her husband. She had not seen her mother for over a year, and so there was a lot of catching up to do. 'What is nuclear waste?' her mother asked, in passing. She explained that the tribal council was making plans to use part of the reservation as a nuclear waste dump. 'I didn't know what nuclear waste was,' Rufina says, 'but it sounded to me like nuclear bombs. I recalled a TV documentary I had seen about toxic waste dumps. The film had made an impression on me because it showed people demonstrating and using the same sort of direct action that we had during the 1960s Civil Rights Movement. So I knew that if nuclear waste was anything like toxic waste, it was very dangerous.' About a week later in Arizona, Rufina had a vision, in which she saw an iridescent liquid oozing out of the mountain from the fresh water springs. The liquid burnt everything it touched and gradually flowed towards the men, women and children of the reservation. 'They were screaming in a way that you do when something totally overwhelms you, a kind of hopelessness. The liquid was going to kill everything. I also saw the faces of the tribal council and three spiritual leaders standing on a hill, helplessly watching the whole thing. I later found out that those spiritual leaders were in favour of the dump. That really shook me.'

Rufina saw her first vision at the age of nine. 'They always start in the same way: I am in an old circular stone tower. There are these crude stone steps that jut out from the inside wall. I climb the steps and as I climb, the light at the top of the tower gets bigger. From the last step, I try to climb out through the hole in the ceiling on to the roof, it is always a struggle. A hooded figure, who never speaks, takes my hand and pulls me up. We then step off the roof of the building into nothingness and the vision begins. To begin with it was like a very clear technicolored dream; the visions were exciting, like movies. I used to see things and go places I have never been in real life. They had a beginning, a middle and an end, and I was always the star! Then gradually I began to understand that they were visions, not dreams. I began to learn how to control them and became so intensely involved with my vision life that I became shy and withdrawn in real life. These visions have come and gone throughout my life but, more or less, they have directed every major decision I've ever made. I have seen happenings in the visions on a number of occasions that have occurred in real life. So the moment I understood the vision with the iridescent liquid, I knew that I had to return to the reservation and try to do something to stop it coming true.'

The first time I saw Wendell Chino was at a Mescalero funeral. The wife of a tribal councillor had died of cancer. In Mescalero they bury the dead in a meadow above town. The cemetery is unkempt. Over the years the gravestones have been swallowed up by the grass and wild flowers as if returning to the earth. About sixty people came to pay their respects. They assembled next to a pick-up truck loaded with garlands of flowers and commandeered by the Gnome Funeral Services from Ruidoso, who made a hash of lowering the coffin into the grave. As soon as the job was done, however, a line formed as each mourner threw a handful of earth on to the coffin. Then the congregation waited in silence while four young men from the reservation finished shovelling the rest of the earth back into the grave. The scene reminded me of the widespread belief that the Native American Indians are the country's first environmentalists, because they have a special relationship with nature, a spiritual respect for the earth. Apaches believe that whatever you take from nature must eventually be returned, and that if you fail to do this an important balance is lost. For this reason some people believe that Native Americans are the only ones responsible enough to be the caretakers of nuclear waste. And yet others are horrified that an Indian tribe even contemplates such a role. At the end of the funeral Wendell Chino stood up and walked over to the pick-up truck where the men from Gnome Funeral Services were getting ready to go. I had been told that Chino was a portly septuagenarian, below average height, so my first impression was of a man both taller and more athletic than I had expected. He wore glasses and his thin white hair was brushed to the back of his head. He seemed to assume responsibility for the proceedings even though it was not a member of his own family that had died. He thanked the men and, as he turned, I caught his eye. He stared at me momentarily, then turned back to the graveside.

Native American tribes have enjoyed a long relationship with the atom. In the 1930s it was discovered that many of the tribal lands were rich in uranium, particularly in the Four Corners area, where Arizona, New Mexico, Utah and Colorado meet. But the stakes were raised immediately after the Second World War, when the Atomic Energy Commission (AEC), then responsible for building America's nuclear stockpile, began to pay a high price for the ore. The AEC set up offices in the four states and offered companies advice on how to prospect and mine uranium. As the sole purchaser of the uranium, it controlled the market and set the price, offering a US$10,000 discovery bonus for high-grade deposits. Mining uranium became patriotic, Americans bought 35,000 Geiger counters in 1953 alone. Hundreds of square kilometres of Indian reservation, previously thought of as worthless, became a potential gold mine. 'Uraniumaires', as uranium millionaires were now dubbed, began to crop up everywhere, and Native American tribes suddenly became indispensable to mining companies because they offered cheap labour as well as invaluable guidance to prospectors. Until questions started being asked about health and safety aspects in the late 1950s and the 1960s, there was no more sophisticated instrument in mining than a stick of dynamite, a wheelbarrow and pickaxe. Miners spent all day, every day, working, eating and drinking in the mines, totally unprotected from the radioactive dust and gases. By the 1970s, a disproportionate number of people on the reservations had lung cancer and other health complaints. One of the first tasks assigned to Robert Alvarez, then a junior staffer for the Democratic senator for New Mexico, was to push a uranium miners' compensation case through the Committee on Atomic Energy on behalf of the miners' widows. 'These tribal leaders and a small group of widows came to my office in Washington,' he told me. 'I thought that it was a pretty open and closed case. The scientific evidence by then was irrefutable and all that really needed to be done was to agree on an amount. But after I filed the case, months went by and nothing happened. I began to get frustrated, and a senior staffer took me aside and told me to forget it. The committee was attempting to push through Congress the new plutonium fast breeder reactor programme, and there was no way that they were going to even pay lip-service to something with these sort of negative implications. "Anyway," he said to me, "they're Indians, they don't have sovereign rights."'

As soon as Rufina Laws returned to the reservation, she formed HANDS (Humans against Nuclear Dumps) and set about educating herself and other members of the reservation about the dangers of nuclear waste. Her campaign was evidently a success, because in January 1995 the tribal council decided to hold a referendum on the future of the waste site. The decision coincided with a move by the Federal Government to discontinue negotiations with Chino, in the light of the beating it was taking in other parts of the United States. The referendum therefore, albeit unnecessary in terms of the reservation's constitution, was intended to settle the matter once and for all, so that negotiations could continue with the power companies.

In the run-up to the vote, Rufina circulated cuttings from environmental magazines, a video made by Nuclear Free America and her own manifesto to every household on the reservation. Everyone waited for polling-day to be announced. Then, one morning, a small byline in the *Ruidoso News* announced that the vote would take place on the last day of January. Rufina rang Silas Cochise, the councillor responsible for the project. 'I was really angry, there wasn't a single radio or TV announcement, just this tiny piece in the paper. I reminded him that they had to notify by letter every member on the reservation of the day and that people were supposed to be properly informed on the subject before they could vote. They were also supposed to have a public meeting about it in advance. He just about had a fit, but that evening I got a letter officially stating the date and calling for a meeting.'

The meeting began with the ten tribal council members one after another telling the audience why they believed in the project. A representative from the nuclear industry was present to answer any questions. Towards the end of the meeting, however, Rufina stood up at the back of the hall and addressed the people. 'I told them that this substance is the most dangerous stuff that has ever been created, it lasts in some instances for 17 million years. One substance has a half-life of 4 million years. We are talking about an eternity, I told them. We cannot vote for this site if we haven't seen a contract. I told them that this is a war, we are in a war for our lands and in all wars there is a war cry. Our war cry has to be, "No contract. Vote No." On polling-day, a record number of voters turned out and rejected the negotiations by a majority of 490 to 362. It was the first time that the reservation had voted against a decision made by their president and their tribal council.

'Right or wrong the people have made their decision,' Chino told the press. 'I don't have a problem with it. I just recognize the fact that the people have shut the door on themselves in not accepting a great opportunity.' But his public expression of fair play and decency disguised pandemonium behind the scenes. When the news of the referendum became public, the nuclear industry was gathered in Las Vegas for a conference on waste transportation. Lila Bird, an environmentalist from Albuquerque who was attending the conference, told a local newspaper, 'You should have seen the stunned look on the faces of the government and industry representatives. They ran to the phones cussing and yelling. All over the world people recognized that if a small group in New Mexico can defeat the nuclear industry, then anyone anywhere can do the same.'

The nuclear industry, however, was not going to be defeated quite so easily. In the next few days the tribe's housing director, Fred Kaydahzinne, and its chief judge, Harrison Toclanny, a man with a glass eye, got to work. The tribal council distanced itself from the pro-nuclear lobby, but using official employees and official vehicles, its members petitioned every household on the reservation. 'If the man in charge of who gets housed on the reservation and who doesn't comes to you and says sign this, do you think that people are going to say no?' asks Joseph Geronimo. Others who were approached claim that they were offered a sum of US$2,000 to sign the petition. 'They came to my house and told me that it was for the sake of my children. Well, you can take that two ways, can't you?' said Netty Fossum, who runs the day-care centre. Within days, the petition with 700 signatories was handed to the tribal council, overturning the previous vote and requesting that negotiations continue with the power companies.

The letter of intent, a non-binding agreement between the tribal council and the power companies, is kept at the Fuel Storage Information Center in Mescalero. Those who wish to read the letter are obliged to sign an agreement stating that they will not reveal its contents to the newspapers. Even so, some of the letter has been blacked out or is missing. 'That is typical,' says Virgil Comanche, who witnessed hydrogen bomb explosions in the 1950s while serving with the US Navy in the Marshall Islands. 'The people on this reservation get most of their information about the decisions the tribal council makes from the newspapers. We feel like cattle in a corral being herded from one end to another.' Now the situation has come to a head. The tribe is divided. For the majority, criticism of the leadership is tacit or confidential. One woman, who had voiced criticism of the plans, pleaded with me not to use her name for fear of losing her job and her home. To the more outspoken, the braves, the situation is now turning into a war. Under his stetson Joseph Geronimo boldly claims that not a single cask of nuclear waste will come on to the reservation. He has vowed to blow up the railway track and to scalp as many heads as necessary to achieve his aims. 'I don't care what they try to do to me. They can take my job and house away, I'll survive by living off the land and I'll strike when they least expect it,' he claims. Most believe that the time has come to elect a new president and tribal council in order to wrestle control of the reservation back into the hands of the tribe members. It is claimed that for three decades a cult of personality (and of fear) has been allowed to develop on the land. 'Owned and operated by the Mescalero Apache Tribe' now rings hollow. Not a single member of the tribe is a manager or assistant manager of the reservation's businesses. No one knows how much those businesses make. 'If you look at the board of directors for the enterprises, you would find that they are all white,' Rufina contends. It is Rufina, however, and not Chino, that has been labelled un-Apache by the tribal council. Her opponents claim that she has returned to the reservation with a white man's view of the world and that she is now trying to convert the Apache spirit. Her family has received death threats and she has been spat at in the street. The continuous hate mail and verbal abuse over the phone have succeeded in driving her out of her mother's house on the reservation, because she feared that the building would be firebombed. She now rents in Ruidoso.

There is nothing tribal about the tribal council headquarters. It occupies a long, single-storey, flat-brick building in the centre of Mescalero. Inside and out, it looks just like any other municipal building, with carpeted corridors, noticeboards and the odd second-rate painting.

I approached the reception desk. 'Is the President in?' I asked.

'Yes,' said the receptionist. 'His office is at the end of the corridor.' I was surprised to find that there was no secretary to sidestep, no assistant to charm, just a door marked, 'WENDELL CHINO – President.' I thought of the brief conversation I'd had on the phone with Chino, and the familiar voices echoed: 'He hates journalists', or

'You'll never get an interview.' As I walked down the corridor past open doors, I saw people look up from their desks and stop conversations. At last I got to 'The Door', knocked, waited for a moment and went in. The President's office was spacious, but hardly presidential in size. In the middle was a large mahogany desk surrounded by matching leather armchairs. To the right of the desk were several glass cabinets full of books. A large houseplant was placed discreetly in the corner. Wendell Chino sat behind his desk. He seemed smaller and older than he had at the funeral. He looked up as I entered the room. In a single breath I explained who I was and asked, 'Can I talk to you about your plans for the MRS?'

He shook his head and said quietly, 'I am busy.'

'Could I talk to you later, perhaps?'

I detected a smile. He was obviously amused by my approach. 'Talk to Silas Cochise, he is the tribal councillor responsible for the MRS.'

'Where do I find him?'

'Ask reception,' he mumbled, lowering his head, and returned to his work. In other words, I was politely but firmly dismissed.

On the way out it occurred to me that the corridor leading to Chino's office served as a kind of funnel through which each call, decision and query passed. I wondered if the absence of office staff exposed the President's inability to delegate, or did he just relish dealing personally with troublesome callers and out-of-towners prying into his affairs.

'The President said I could talk to Mr Cochise,' I told Silas Cochise's secretary.

'He did?' It turned out Cochise wasn't in his office but in any case the secretary was unimpressed. When her boss returned I explained again that Wendell Chino had sanctioned the interview.

'He did? Well if the Chief says I can speak with you, you'd better come in.'

Silas Cochise, who is the great-grandson of the chief of the Chihuahuan tribe, rehearsed the arguments for the waste site. He explained the tribal council's commitment to long-term employment, education, improved housing and health facilities. He spoke of having seen a number of waste sites across the United States and considered them perfectly safe. He mentioned that the additional business spin-offs from the site would ultimately benefit the community. He told me that the tribal council felt that someone had to take the lead in dealing with the nuclear-waste problem because it was a universal issue. He expounded the racism argument that it's fine for New Mexico to host government facilities like WIPP and Los Alamos, but all of a sudden unthinkable for Mescalero Apaches to store radioactive waste. He told me that the tribe was sick and tired of holding its hands out to the Federal Government and getting nothing. 'We are investing into a sustainable future for our tribe, one in which we can hold on to our tribal traditions and culture.' Tribal traditions were slipping away, he claimed, as the younger generation, attending state schools outside the reservation, became more and more influenced by white culture. 'We never used to have a graffiti problem, nor did we have a problem with marijuana and alcohol on the reservation. Many of the young don't know how to speak Apache or sing the traditional songs. Others don't understand the meaning of the ceremonies. This facility gives us an opportunity to build our own schools so we can teach these traditions.' As to opposition in the community, Cochise suggested that he could count the activists on one hand. 'Five people,' he kept saying, 'only five people. We are doing our best to educate the people about the facility, but there are always some who don't want to be educated. Do they want us to return to living in tepees?'

The idea that it's nuclear waste and gambling that represent the tribe's sustainable future, while its downfall are a mixture of graffiti and alcohol, seemed to me an odd way of looking at things. 'We're not part of America,' Rufina Laws told me. 'So why are we behaving as if we are? Native Americans have this schizophrenic character. One minute we are Indians, the next we are Americans, depending upon whether we are on or off the reservation. Wendell Chino has understood this very well. Ultimately we have to decide who we want to be.'

'People have allowed this situation to go on for too long,' Netty Fossum explains. 'In many ways it is our own fault. My worry is that if we can't deal with this problem using the ballot-box then people will resort to the gun, and blood will be shed.'

Just before I left the reservation Rufina Laws became a candidate in the following presidential election. Her message to the electorate is clear: the reservation will never house nuclear waste as long as she lives. Recently she had another vision in which the trees came to her and told her that they supported her and that she must continue with her work. In most presidential races such talk would no doubt signal the end of a political career. But things are different on the Mescalero reservation and if anything can bring them victory, it is Rufina's fierce commitment to the spirit of the Apache nation and her sense of its past.

6. *Nuclear Nomads*

In May 1995 the people of the Bikini Atoll voted to oppose a plan to dump nuclear waste on their island. Shortly afterwards, I received a letter from Dr Steven Simon, the Director of Radiological Studies for the Marshall Islands, who wrote, 'The option has been rescinded, partly due to my testimony, thoughtful thinking by the Bikinians and some concentrated effort by the liaison officer.' The liaison officer, Jack Niedenthal, an American married to a Bikini islander, informed me that the issue of nuclear waste was so politically unsettling that many islanders were afraid to talk openly about it. Niedenthal had threatened to resign if they voted in favour of the plan.

The Bikini Atoll, one of the larger of the Marshall Islands, is situated 4,000 km south-west of Hawaii. The thirty atolls of the Marshalls form part of a vast archipelago in the Pacific known as Micronesia. From the air they resemble a handful of coral necklaces dropped at random on to the blue velvet waters of the ocean. Each glistening atoll is a string of reefs and islands that encapsulate a shallow lagoon. These idyllic natural harbours have provided shelter and rich fishing grounds to the islanders for centuries.

Towards the end of the Second World War the Marshall Islands were liberated from Japanese occupation by the Americans. Yet only six months after VJ Day, the 167 people living on Bikini were told by the US Navy to leave their homes and go to the atoll of Rongerik, 200 km east of Bikini. In pursuit of world peace, the island was being commandeered as the laboratory for the world's first series of atomic tests. 'An American came to Bikini. He said that he was the most powerful man in the world. He said he wanted to drop a bomb on Bikini. He said America wanted to use Bikini and that we would have to leave,' Kilon Baunu, one of the evacuees, told Robert Stone, the maker of the film *Radio Bikini*. 'It is difficult for me to express how sad

I was as we were leaving. We looked back and saw them burning all our houses. They burned everything.'

'When the Bikinians were moved, it was something that really shattered their hearts,' Tomaki Juda, youngest son of the chief islander, explained. 'But they made promises, and we held on to those. The United States was the leader of the world. Did we have a choice?' The US Navy did not anticipate any problems moving the islanders; one atoll, after all, looked much the same as another. Government newsreel reported at the time that the Bikinians 'are a nomadic group and are well pleased with the Yanks, who are going to add a little variety to their lives.' The islanders, who up until that point had never been nomadic, were about to become the world's first 'nuclear' nomads.

As the Bikini islanders were moved off their island in March 1946, ninety-five ships damaged or captured in the Second World War were dragged into the 43-kilometre lagoon as ground-zero fodder. The vessels were laden with tanks, planes, ammunition and 400 tethered goats and pigs, in an experiment to assess the power and capabilities of the atom bomb. On 1 July 1946, the Americans dropped *Able*, the first in the series of explosions known as Operation Crossroads. The 23-kilotonne bomb dropped by the B-29 bomber *Dave's Dream* was detonated above the lagoon. Thirty kilometres outside the atoll, the explosion was watched by over 42,000 military personnel, heads of state, politicians and journalists. As the seconds were counted down the world listened in silence to a live radio broadcast. It was like a 'huge, giant firecracker', said a witness after the explosion, 'a setting sun', recalled another. In truth, the blast was disappointing. As the mushroom cloud cleared it revealed that the majority of the ships, albeit burnt and twisted, were still afloat. *The New York Times* referred to the bomb as a 'dud'.

Three weeks later *Baker* was detonated in the lagoon at a depth of 27 metres, in the world's first underwater test. The blast burst through the surface of the water with such force that it lifted the ships in its path and tossed them aside. A column of water a couple of kilometres wide rose higher and higher, sucking up and pulverizing the contents of the lagoon, while spitting radioactive coral and sand across the atoll. From the base of the explosion burst a huge radioactive wave of vapour and particles that rolled over the surrounding islands. Within a couple of hours of the blast the radiation-monitoring crews went into the lagoon to take readings, but quickly retreated. Twelve hours later it was still too 'hot' to enter. Over the next two weeks the wreckage was explored, but it became clear that the contamination was still too severe to work with. Moreover, it was being spread by radiologists and military personnel who had climbed aboard the ships still afloat in the water. On 10 August Operation Crossroads was abandoned, and the entire military circus packed up and went home. The world's press reported the fact that only sixteen ships had been sunk by the explosions, but they missed the real story. The story was radioactive fallout. The eminent scientist Glenn Seaborg observed that the *Baker* test was 'the world's first nuclear disaster'.

After Operation Crossroads, normal life resumed in the Marshall Islands, and the evacuees from Enewetak, Wotho and Rongelap returned home. The Bikinians expected to be next to return, but as the weeks and months went by with no sign of repatriation, the tension between the US Navy and the exiles began to rise. The Navy's explanations of radiation, or 'poison', as the Marshallese call it, were met with a blank stare. Likewise, the Navy did not appear to understand the islanders' longing to return to their homes. Vice-Admiral Blandy, in charge of Operation Crossroads, summed up the Navy's feelings, 'We wish to acquire . . . a few miserable islands of insignificant economic value, but won with the precious blood of America's finest sons, to use as future operating bases. All that can be raised on most of these islands is a few coconuts, a little taro and a strong desire to be somewhere else.' Over the next few months the Bikini islanders' quality of life on Rongerik Atoll declined steadily. During the winter months there was a severe shortage of food and the Bikinians began to starve. Rongerik was not only smaller than Bikini, but a high proportion of the fish that frequented the waters were poisonous. To add to the problems, the Bikinians felt that Rongerik was inhabited by evil spirits. The Navy dismissed these concerns as a mixture of superstition and homesickness, but conscious of the threat of bad publicity they established a committee to explore the idea of resettling the islanders elsewhere. They decided that the uninhabited Kili Atoll, some 650 km to the south, or the populated atolls of Ujae and Wotho might be suitable alternatives. In July 1947 the Bikinians, starving, banished and now desperate, announced in writing that they would move to Ujae, but added that they would have preferred the unpopulated island of Kili, if it had had a lagoon and had not been such a distance from their home. Their confused message to the US Navy ended with the pathetic submission, 'We will do whatever you say.' After further bewildering discussions, they decided to stay where they were. The following year an anthropologist visited Rongerik and reported that a fire had destroyed one third of the trees and that the now wretched islanders were forced to eat the young palms and the poisoned fish. News that the Bikinians had not been adequately compensated spread quickly and the US Government was accused of atomic colonialism. In response the Navy attempted again to move the islanders to yet another atoll, Ujelang. In a supreme piece of bureaucratic bungling it transpired that the Atomic Energy Commission (AEC) had already promised the island to the people of Enewetak, in return for the use of their atoll in further atomic testing. The Bikinians' disorientation deepened as the US Government continued to shuffle and reshuffle the people of the Marshall Islands. Eventually, the Bikinians were moved to a temporary camp on Kwajalein Atoll, a US military airbase, whilst they explored the possibility once more of moving to Kili. The island had many disadvantages, it was smaller than Bikini, with no lagoon, and had little protection from strong winds. But its one advantage was that it was not subject to the Marshallese system of taxation known as the *iroij lablab*. Based on this advantage, the Bikini islanders moved to Kili, where the majority remain to this day.

Twenty-three atomic bombs were detonated on Bikini before 1958 when Nevada became America's only atomic test site. But it was the *Bravo* test in 1954 that sealed the fate of the Bikinians. A 15-megatonne hydrogen bomb was detonated on the island of Nam in the north-west of the atoll. The explosion, 750 times more powerful than the bomb dropped on Hiroshima, vaporized three small adjacent islands. As the force of the detonation unfolded, its radioactive fallout not only smothered the Bikini Atoll, but showered the populated islands of Rongelap and Uterik. The event is regarded as the worst single incident of fallout exposure in the history of America's atomic testing. In all, sixty-seven nuclear explosions were detonated on Enewetak and Bikini, contaminating a number of islands that are now permanently uninhabitable because of the high levels of plutonium and caesium in the soil. Recent de-classification of documents at the Department of Energy has revealed that the US Government was well aware that the islands were far more contaminated than was ever admitted to or compensated for. Up to now it has only provided the Marshall Islands with a total of US$150 million. The Bikini islanders secured US$75 million, which leaves the other contaminated atolls seriously underfunded. The question facing the Marshallese was whether to abandon these islands forever or attempt to clean them up. They have now opted for the latter. It is estimated that the clean-up of each island will come to US$350 million. Attempts by the Marshallese to get the US to accept responsibility and release more funds have so far failed. As a result they have turned to the international community to find ways of generating money for the clean-up.

Of all the Marshall Islanders, the Bikinians have been affected most dramatically by the last fifty years. Decades of broken promises, starvation and a culture of dependency have irreparably damaged the community and its traditions. The few elders who can remember how things used to be before 1946 complain that ancient fishing skills, which have been handed down through the generations, are being forgotten. Over the years, the recognition of their struggle, the protection of their interests and the securing of compensation have primarily been the achievements of two men: Jack Niedenthal and the Bikini lawyer Jonathan Weisgall. In 1984, at the age of twenty-six, Niedenthal started working in the Marshall Islands as a Peace Corps volunteer teacher. After three years in the job he was given the role of liaison officer for the Bikinians. He is now responsible for their day to day affairs and, being fluent in Marshallese, he is their much-needed link with the outside world. In 1968 President Johnson announced that Bikini was radiologically safe. The following year, the islanders returned to their atoll but were concerned about the truth of the

President's pronouncement. By 1975 they were looking for legal representation and were put in touch with Weisgall, then a young lawyer with the Washington DC practice Covington & Burling. After assessing the facts, it was decided to file a lawsuit which asked the court to demand that the Government do a complete radiological study of the island. The subsequent study exposed the fact that the Atomic Energy Commission's calculations of what was 'safe' were seriously flawed. Tests showed that the islanders had far exceeded the maximum permissible dose and had significant amounts of caesium in their bodies. In 1978 the islanders were evacuated for a second time from their atoll and returned to Kili.

In addition to US$75 million originally paid in compensation by the US Government, the Bikinians obtained a further US$90 million by 1992 as a result of legal action. The annual budget of US$8 million a year ensures that the growing Bikini population, now estimated at 2,000, will never starve again. Today life on Kili is dominated by cars, video shops, a sports centre, laundromats and casinos. Each house now has air-conditioning, and the locals eat imported hot dogs and tinned Alaskan salmon. According to Jack Niedenthal, this sudden explosion of money has resulted in an overwhelming sense of greed. 'The Bikinians are very materialistic,' he laments. 'Money is God to them and they don't care where it comes from.'

Now that the Department of Energy has started work on the environmental restoration of the atoll, the Bikinians are once again planning to go home. But they are faced with a serious question: What are they returning to? Their dilemma was depicted in an article for *The New York Times* by Jeffrey Davis, a journalist who met the islanders at one of their quarterly policy meetings held in Las Vegas. The location is a favourite of the Bikini council. 'If we could make Bikini into something like Vegas, we'd know we were really headed somewhere. That would be a dream come true,' Jamore Aitap told Davis. The questions of the atoll's future and of how to provide enough funding for its complete clean-up has divided Niedenthal and Weisgall. Niedenthal's cautious idealism and his concern for maintaining what is left of the Bikini culture is at odds with the other's functional approach. His proposals include converting the atoll into an exclusive holiday resort, complete with diving expeditions to explore the rusting battleships at the bottom of the lagoon and adventure weekends for those interested in self-sufficiency and encounters with tiger sharks. However, most of the schemes pale besides the millions of dollars promised in exchange for a nuclear waste repository.

In January 1988, a month after the Marshall Islands were accepted as a possible location for America's spent nuclear fuel, Weisgall approached Waste Management Inc., one of the largest waste companies in the US, to assess the feasibility of storing nuclear waste on Bikini. As Weisgall explained to *The New York Times*, Bikini already has a radioactive dump site where the *Bravo* hydrogen bomb was exploded. 'Why not charge rent? It would be of enormous benefit to the Bikinians, not to mention the employment opportunities.' As word travelled that the Bikinians were interested in the possibility of using part of the atoll as a waste site, Weisgall was approached by an international consultancy called Pan Pacific Consolidated Ltd (PPC). PPC had the most honourable reasons for declaring their interest in developing a suitable interim site for the storage of spent nuclear fuel. In a letter to Weisgall, Admiral Daniel J. Murphy, a retired US Naval Officer and chairman of the company, wrote, 'Jonathan, before we go any further, it is vital that you understand the reason for our efforts. It is our primary objective to contribute to the total cessation of reprocessing. We know that if such a facility as the one proposed by PPC were in existence today, there would be global pressure both politically and economically upon British Nuclear Fuels Ltd and Cogema [French nuclear fuel authority] to halt reprocessing operations. Without the proposed PPC facility, the only alternative is to reprocess and subsequently manufacture plutonium. PCC is not interested in nuclear waste disposal; but we are interested in temporary storage.' The letter goes on to say that 'in offering the first truly global non-proliferation solution of this century, PPC and the Bikinians would be deserving of the Nobel Peace Prize.' The Admiral's letter ends with the details of the proposal, in which the various phases of the project are explained. The development phase has an estimated cost of US$10 million. A non-refundable payment of US$10 million would be forthcoming on completion of the first phase. Phase two, the building of the storage site, containers and ships required to do the job, along with the government relations and public relations work, is estimated to cost a further US$9 billion and will take an ambitious thirty-six months.

Admiral Daniel J. Murphy has friends in high places. Much of his career was spent working in Washington at the Pentagon for the CIA and as Chief of Staff and top foreign-policy expert for Vice-President George Bush. Having retired from office in 1985, he worked as a lobbyist using his influence and connections. At the end of 1994, during meetings with the Department of Energy in San Francisco which lasted several days, the Bikini council were introduced to Alex Copson, the president of PPC. 'He was this brash British guy with his slicked-back hair,' says Niedenthal. 'He'd seen some film of a train piling into a nuclear fuel cask and said it was perfectly OK. We didn't like him. He knew I was the stumbling-block. He also had this financial guy with him whose claim to fame is being responsible for the come-back of Buffalo Springfield!' The meeting was not a success. But although the islanders had been unimpressed with Copson, they were still keen on the non-refundable US$10 million. At the beginning of the following year they voted to look further into the matter. 'It seemed ludicrous to me that the Bikinians, who had convinced the world that they loved their island, were all of a sudden considering turning it into a nuclear dumping ground for money,' says Niedenthal.

The issue of nuclear waste storage in the Marshall Islands was introduced by President Amata Kabua. In the late 1970s President Kabua went to Tokyo to discuss with a Japanese delegation the possibility of storing nuclear waste on one of the uninhabited Marshall Islands. Even then, the issue of nuclear waste was considered a sensitive topic. So in order not to draw attention to himself, President Kabua travelled under a different identity. His whereabouts were leaked to the press, however, and a flurry of newspapers carried reports of the trip. On the President's return, according to Niedenthal, the islands' council 'jumped on his head'. The issue of nuclear waste didn't resurface until a decade later. In 1987 President Kabua approached the US Government offering the islands and, in a last minute amendment, due to some energetic lobbying, the name of the Marshall Islands was included in the Nuclear Waste Policy Act Amendment, along with the fifty states and the Native American reservations.

Admiral Murphy had also been in contact with President Kabua to explain PPC's proposal, again emphasizing the benefit to world peace and harmony. The letter to Kabua, however, unlike the one to Weisgall, contains no mention of the storage facility being 'temporary'. The attraction of the proposal is that the Admiral's contacts 'could provide a favourable and successful outcome' in convincing the US Government to pay more attention to the Marshall Islands' wish to be considered as a suitable waste location. The letter concludes, 'The major burden rests on the acceptability and recognized capability of those who would perform the collection, containment and transport aspects of this project. I am proud to inform you that PPC and its partner companies are unique in providing that part of the equation.' President Kabua has left Admiral Murphy's letter unanswered, and when I enquired, a source close to the Marshall Islands Government refused to discuss anything concerning the company, Pan Pacific Consolidated Ltd. However, the Marshall Islands and an as yet unspecified South-east Asian country are currently negotiating to develop a low-level waste site on one of the contaminated islands of Enewetak. The operation will be run by an American firm (to be appointed) which will pay rent to the Marshallese Government. I was told that the agreement 'will provide additional funds that will be meaningful in relation to the scope and nature of the problem', meaning that millions of dollars will be paid annually to go towards the clean-up of the islands. The site would also have the additional advantage of providing housing for the islands' own radioactive waste. In an irony unique to the nuclear age, the curse is also the cure.

Pan Pacific Consolidated no longer exists. But Admiral Murphy and Alex Copson have a new company, US Fuel and Security, and it is setting up a factory on the banks of the Mississippi, in the town of Helena, Arkansas, where it will be manufacturing dual-use storage and shipment nuclear casks.

As Dr Simon stated in his letter, after some 'thoughtful thinking' the Bikinians, independently of the Marshall Islands, voted for a future of diving expeditions and élite tourist resorts.

The United States National Oceanic and Atmospheric Administration presented a paper at the Earth Summit in 1992 that assessed the possible effects of global warming. The study concluded that if, as expected, the polar ice-caps start to melt, the ocean levels will rise by as much as a metre in the next century. As a consequence, certain islands in the Pacific and Indian oceans will be flooded, and the entire population of the Marshall Islands will have to be relocated.

PART THREE

7. *The River*

The river Techa is not a deep river – you can wade across it at any point. Most of the winter the river is hidden from view by snow, but in the short, hot summer it abounds with plant life and fish. The Techa was an important river in the past. It was the life blood of the forty or so villages through which it flows. It also played a part in the Cold War.

The secret atomic city of Chelyabinsk 65 was built on the banks of the river Techa in 1948. It was dreamed up by Stalin, partly in response to the bombing of Hiroshima, in order to manufacture plutonium for the 'Red Bomb'. Between the end of the Second World War and 1949 when Russia exploded its first atomic weapon, eleven secret cities were built throughout the Soviet Union. The cities were 'secret', insofar as they were classified by number and granted top national security status – in other words, it was forbidden by law for anybody who had visited the cities to speak about them in public. Even now, with many sites open to foreign visitors since the end of the Cold War, no maps reveal their precise locations.

Chelyabinsk has the dubious reputation in the West for being 'the most polluted spot on earth', a reputation that the local people seem happy to endorse – they have heard about the large amounts of money being poured into Chernobyl. My interpreter, for example, observed cheerfully that the problem in Chelyabinsk was ten times worse, but whereas the international media reported the accident in the Ukraine within a matter of days, it took forty years for the environmental disasters at Chelyabinsk to be made public.

As soon as my plane landed at Chelyabinsk airport, a reception committee marched onto the tarmac with the express purpose of putting me straight back on the plane unless my papers were in order. Money, passports, Western cigarettes were handed over until it was decided that my travel documents were up to scratch – and the atmosphere improved. I was introduced to my guide, a city apparatchik with a drink problem, who ushered me into the VIP lounge for a glass of vodka and a meal. Afterwards I was driven to my hotel, the newest in Chelyabinsk and therefore regarded with much pride by the local people. The hotel's marbled but unfurnished lobby had become a fashionable place to hang out in and to watch the satellite TV channels that were showing day and night on an old television in the corner. Upstairs there was evidence of unfinished maintenance work: handfuls of coloured wire were suspended from the ceiling; aluminium ladders were stacked up in the corridor. It was clear that I was the first person ever to occupy my room. There was no hot water and the water that dribbled out of the taps was brown. I remembered the statistics I had read on the flight about the water in Chelyabinsk containing five to twenty times the permissible levels of iron, thirty to sixty times the levels of zinc, and forty to sixty times the levels of copper.

Between 1948 and 1951 high-level waste from Chelyabinsk 65 was pumped into the river Techa. The practice was stopped when radioactive particles were detected at the mouth of the river Ob, in the Arctic Circle, approximately 1,500 km north. The Techa runs 240 km before it becomes the Iset, which flows north into the Tobol, which in turn becomes the Irtysh and, eventually, the Ob. It was feared that the radioactive particles would give away the location of the secret plant. Even though the high-level waste was diverted into nearby lakes, low-level discharges continued until 1956. Being neither a large nor a powerful river, the Techa washed only 1 per cent of the discharges further than 35 km from the plant. In some places, unusually large concentrations of strontium-90, caesium-137 and plutonium-239 were detected in the river's bed and banks. The evacuation of the villages along the Techa started in 1953, not long after a team of biophysicists from Moscow raised the alarm. For the next seven years 7,500 people from twenty-two villages were relocated, but it was already too late; it was estimated that the average yearly dose received by these people ranged between 3.6 and 140 rems, and many individual doses were much higher (the acceptable yearly dose for the public in the UK is 0.5 rems). The villagers of Metlino, only 7 km from the plant, received the highest doses. But if you were to go there today, you would find that the village is underwater. Russian officials flooded the area when they dammed the river in an attempt to halt the flow of nuclides. All that remains visible of Metlino is the top of the church steeple, which breaks the surface of the water in the middle of a vast reservoir.

By 1961 every village within 40 km of the plant had been relocated – except for one. To this day, no one is clear why the people of Muslimova were left behind. Many of them feel that they were unwitting guinea-pigs in a grotesque experiment to identify the long-term effects of low-level radiation. Some argue that the town was important strategically, as the railway to the nuclear plant passed through it. Others see it as a racially motivated decision: Muslimova, as its name implies, is Tartar.

The dirt roads in the centre of town are lined with oddly suburban, multi-coloured bungalows. Each tiny front garden is fenced, with a small gate and a path leading to the door. The mosque, a corrugated-iron building painted blue and yellow, is situated on the outskirts of the town, which is really a large village. People walk up and down with bundles of firewood on their backs, others pass by on horse and cart. As we crossed the concrete bridge in the middle of the town, people stopped in the road and stared. I could see them thinking, 'Yet more experts come to study us.' A sign innocently proclaimed, Techa.

Instead of being relocated, the Muslimovans were instructed not to use the 'dirty' river. The town council dug wells for use as an alternative water-supply. Houses near to the river were demolished. Barbed wire was erected along its banks. But the river did not look 'dirty'. It looked as clear and fresh as it had ever been. Ducks continued to swim up and down, and the plant life appeared to be normal. Defying the council diktat, farmers cut down the barbed wire and used the river to water their animals. The women of the town continued to wash clothes and dishes in it. Children fished and played in it. By 1956 a river patrol was employed to repair the barbed wire, and people caught using the river were fined. Each patrolman, however, signed the Official Secrets Act forbidding him to talk about the danger in any detail. As a result, the safety measure had little effect. 'We would wait until the patrolmen were out of sight and then use the river,' a Muslimovan, who grew up here in the 1950s, told me. 'We used the river to fish in. Sometimes we would catch a dead pike that was floating down stream and take it home to eat.'

One afternoon I was taken to the river-bank and shown the radiation levels. It was a hot day and the water looked fresh and inviting. On the far side cows were drinking. My escort, a town councillor, went to the water's edge to get the reading. He cleared a small patch of weeds and placed the Geiger counter in the wet mud. Its needle immediately jumped to life and swung to the far right of the dial. An old woman stopped to watch. As soon as she realized what was going on, she began to shout at us in Russian. My host smiled, as if to say, 'She's mad,' and the driver told her to get lost.

At the nearby hospital, the walls and floors had been painted an acid yellow. The colour scheme made the place feel small and claustrophobic. Dr Gulfrida Galimova, dressed in a starched white uniform, sat behind a desk surrounded by her patients' files. She was unwilling to talk to me in front of the apparatchiks, but eventually we managed to get rid of them by insisting that they go outside if they wanted to smoke. Then she began to tell me about her life in Muslimova, where she came to work full time in 1981. Up until her appointment the town had had no resident doctor even though, as she soon found out, many of the women were anaemic and 10 per cent of childbirths were premature. Dr Galimova questioned the statistics, but was told by the Urals Research Centre that it was due to the Tartar way of life: a lack of good food and too much alcohol. And yet the official version contradicted what the doctor could see with her own eyes: all the townsfolk grow fruit and vegetables in their gardens. Having heard the rumours circulating in the area, she enquired if radiation might be the cause of bad health. She was informed that the releases had occurred so long ago that any significant levels of radiation would have dissipated by now. It was implied that she was not a good doctor. In 1986, however, Dr Galimova attended a conference about the health hazards resulting from the accident at Chernobyl. The symptoms highlighted were exactly those she was seeing in her town: low blood-cell count, constant headaches, pain in the joints, stomach problems, pain in the spine, chronic fatigue, sterility, asthma, arthritis and cancers.

I could hear raised voices outside the door. A group of women were demanding to speak to me and they were being jostled by the apparatchiks smoking outside. The protesters wanted to tell me that they weren't getting any of the compensation promised to them by Boris Yeltsin, and that no one was doing anything about their children. They wanted to know why they had not been relocated and why they had been lied to for forty years. 'We knew that the other villages had been moved, but we didn't dare ask why. In those days you did not ask questions,' Sharia Khmetova told me. 'You did what you were told to do, it was too dangerous to do otherwise.' Throughout the 1950s and 1960s, Sharia explained, she had been totally unaware that her husband worked for the atomic project. He was recruited by biophysicists at the nuclear plant to collect samples of mud and vegetation from the river-bank in glass containers. Once a week the samples were collected at a prearranged location. It seems that the biophysicists monitoring the radiation feared that their presence on a regular basis might arouse suspicion in the community. Sharia only discovered this when her husband died and she was asked to continue his work.

The people of Muslimova trust nobody. They distrust the radiologists from Chelyabinsk and they distrust the Western radiologists who are currently studying the area. They dislike officials – and to some extent their feelings are justified. After all, the full horror of their situation was only brought home to them in 1989, when a television marathon raising money for cancer patients included a short film about the plight of a Techa woman who had died of cancer. The film showed a scientist taking radiation measurements on the banks of the river, *their* river.

Dr Galimova refers patients with serious illnesses to the Urals Research Centre in Chelyabinsk (the very same centre that once questioned her professional ability). The centre is unique; its basement houses a comprehensive library of health records of those people most seriously irradiated by the releases from the nuclear plant. To the epidemiologist, the archive is an invaluable resource in the study of long-term effects of low-level radiation. Thousands of cases dating back to 1951 are kept in brown paper files in cabinets and shelves in half a dozen rooms. Each file has a patient number and a series of markings on the outside. Two red stripes means that the patient has leukaemia, a large black dot indicates that the patient received a large dose. This unique collection of data gets scant attention from Moscow and the West. It has never been copied and there is not a single fire door or extinguisher in any of the rooms. (In contrast, health records from the Marshall Islands, where the US Government did much of its atomic testing, and the Hiroshima/Nagasaki statistics, exist in several copies and the originals are heavily protected.)

Mira Kossenko, head of epidemiology at the Urals Research Centre, has such inadequate funding that she finds it possible to undertake follow-up studies only of Muslimova and the surrounding area. She has to overlook the victims of an explosion at Kyshtym in 1957 or the contamination from Lake Karachay. Dr Kossenko has worked at the centre since 1967. She was twenty-nine when she got the job. On her first day she was instructed to sign the Official Secrets Act, forbidding her to acknowledge the existence of radiation sickness. She was permitted to tell her patients only that they were suffering from 'ABC disease' or 'weakened vegetative syndrome'. The permanent presence of armed guards in the hospital constantly reminded the staff of the seriousness of the oath. 'Most of the time the patients were poorly educated peasants, they never asked what was wrong with them,' adds Dr Kossenko. 'People were not encouraged to ask questions. We were all afraid. But we did try to do everything we could for those people. That is the tragedy of my country and my people.'

Nowadays the armed guards have gone and although Mira Kossenko is permitted to tell her patients what is wrong with them, she has other problems. Three years ago Dr Kossenko and her team were able to set up a dose reconstruction study with funding from Moscow. By August 1994, however, the money had dried up and patients were having to fund their own trips to the clinic. 'We are losing the population that we have been studying since the 1950s,' Kossenko explained. 'Nobody seems to care about that. All we have had from the West are promises.' Due to a lack of funding, the hospital, with a capacity for fifty beds, can provide care for only fifteen patients. On the top floor of the clinic a corridor stretches the length of eight wards on both sides. Most of the wards are now empty, with beds stripped and walls left bare. Medication and blood transfusions at the clinic are limited. There is no money to pay blood donors. All operating is referred to the main Chelyabinsk hospital. The equipment, though functional, is very old and, in any case, it amounts to no more than an electrocardiograph and a device for taking biopsies from bone marrow. Next door, a sterile ward for people with myelomas is equipped with an infrared light and a fridge.

The first patient I spoke to had leukaemia. He was born in Muslimova in 1944. He recalled spending most of his childhood playing in the river. He started to get ill in his early twenties, when he left the army. He began to feel tired all the time and thought that it was as a result of his service in the Rocket Corps. Of course, there were some rumours about the Techa being polluted, but few people took any notice: the river *looked* clean. Ironically, lying in the next bed was a river patrolman. In 1956, when he had started working at Chelyabinsk 65 at the age of twenty-two, he had been told not to go near the river because it was polluted with radioactive waste. As a river patrolman, he told me, he was constantly telling people to get out of the water because it was dirty. He was aware that none of the villagers believed him, so he had to resort to making on the spot fines of anything between ten and twenty roubles for children, and fifty roubles for adults. His own children continued to swim in the river, however, even though he punished them for doing so. 'What can you do?' he asked. In 1986, upon his retirement, he went for a medical check, and it was discovered that he had heart disease, a kidney infection and bronchitis.

Over the years some 76 million cubic metres of high-level waste were released into the Techa. The total radioactivity is estimated officially to be in the region of 2.75 million curies. Around 124,000 people were exposed and, according to official statistics, 28,000 of them received a dose of 'medical consequence'. A report by the Russian Institute of Biophysics, commonly known as the blue book, goes on to state that more than 8,000 patients have died so far. It is too early to say, of course, but Dr Kossenko and her team believe that the people living on the Techa are twice as likely as the general population to get leukaemia. Skin cancers have increased fourfold since the 1960s. Cancers have increased by 21 per cent in the last decade. Birth defects have increased by 25 per cent. According to Dr Galimova, the 'river disease', as it is called in Muslimova, has claimed a life in every family. 'It is vital that these people are relocated,' she told me. 'Those who were relocated are getting healthier, generation by generation. But the community here is not getting better.' Mira Kossenko agrees that it is vital to separate the patients from the source of the radiation immediately. The officials are less convinced, however; they say that there is no need for evacuation as long as the villagers of Muslimova do not break the rules by breeding geese or ducks, by letting their cows drink from the river or by fishing in it. Some point to the Russian village of Brodokalmak as proof that it is the Tartar lifestyle, and not radiation, that is responsible for the sickness and disease in Muslimova. The population of Brodokalmak was not evacuated and their health record is superior, but the village is located much further down the river.

Forty years ago a dam built on the Techa, up river from Muslimova, formed a reservoir to hold the contaminated waters. A few years ago the water rose to only 20 cm below the top of the reservoir, so the wall of the reservoir was raised by a metre as a precaution. But by August 1994, the gap was closing again. Four hundred thousand cubic metres of radioactive water is held in the reservoir. 'We could be facing a radioactive flood in Muslimova,' I was told by a town councillor. 'It depends on the weather.'

Russia has the largest nuclear submarine fleet in the world, but an estimated 126 of its submarines now urgently require decommissioning. The number is expected to increase to 200 by the end of the century. The United States is currently decommissioning 85 vessels and Britain 10. Since 1990 the Russians have used spy satellites to monitor the progress of US arms limitations. By hovering over the Hanford nuclear reservation where the US Navy dumps its nuclear submarine reactor compartments, the Russians can tell exactly how many vessels have been decommissioned. The reactor units line a giant trench the size of three football pitches in the northern part of the Hanford site. The trench is deliberately left uncovered for the benefit of the Russian satellites. But as you drive past, you are asked politely by a DoE guide to avert your eyes. In a way this bizarre request only serves to emphasize the fact that during the Cold War the two superpowers knew more about each other's capabilities and operations than their own respective populations did.

On the other side of the world, US spy satellites monitor the number of Russian nuclear submarines being retired from use in the Arctic coastal areas of the Kola Peninsula. The Barents Sea is home to Russia's northern fleet of attack submarines. Having been enlarged several hundred times, the satellite images show small, dark, pencil shapes clustered in the craggy fiords. Approximately 100 retired and disused submarines are now permanently moored in this frozen wilderness. Their dark, shadowy hulls lie rigid, suspended in ice, in a frozen graveyard of rusting firepower. Their future is uncertain. There is neither the finance, the expertise nor the collective will at present to decommission the fleet. Today, foreign government officials, Western consultants and environmentalists are escorted around the naval bases to assess the environmental damage, safety and disarmament problems.

The beaches at Murmansk, in the northern part of the Kola Peninsula, are littered with Cold War flotsam and jetsam. These rusting oily hulks of metal are pumped with compressed air to keep them afloat and nursed by a skeleton crew – it is a hospital ward for the terminally sick. Most of the submarines still contain approximately 500 spent fuel rods. It is estimated that in a couple of years the number of spent fuel rods requiring immediate removal from the disused submarines along the Kola Peninsula will rise to 100,000. But the standard equipment for removing fuel rods from reactors is crude; technical resources and planning are equally poor. Proliferation is yet another concern. Few of the naval bases are adequately protected. In 1994 three Russian naval officers were arrested for stealing 4.5 kg of nuclear fuel worth US$700,000.

The removal of spent fuel in these inadequate conditions is

especially dangerous. On 10 August 1985 the reactor of a Russian nuclear submarine exploded while it was being refuelled at Dunay submarine facility in Chazma Bay, just east of Vladivostock. The explosion occurred when the control elements of the reactor core were inadvertently moved as the reactor lid was being lifted. News of the accident did not begin to leak out until five years later. The trade union daily newspaper, *Trud*, published an eyewitness account, 'The reactor cover was slowly creeping upwards when it suddenly went askew, knocking against the lattice. The reaction started. High pressure, super heated steam broke loose from the reactor depths, hitting the cover with great force. The ship repair yards shuddered from a powerful explosion. Everyone rushed to see what had happened. What they saw were flames and brown fumes bursting from gaping holes in the crippled sub. The one-tonne reactor cover was thrown about 100 metres by the explosion, almost to the other side of the bay. Ten people who were on board the sub were killed.' Three hours later the accident radiation readings were off the scale – Geiger counters register a maximum of 600 roentgens. It is estimated that a third of the entire military area was contaminated. Decommissioning the vessels of the northern fleet will eventually be carried out in dry docks at Severodvinsk several hundred kilometres to the south of Murmansk. The process involves cutting out and sealing the submarine reactor. The work is slow and costly; in 1994 only two submarines were in the process of being decommissioned. Electricity failure is common at the docks and the workforce is not paid for months. The future of this radioactive debris remains unresolved by the Russian authorities; railway-yards and dockyards overflow with old reactors and spent fuel, and the northern fleet lies rotting in frozen waters.

Novaya Zemlya is an island in the Arctic Circle several hundred kilometres north-west of the Kola Peninsula. It was a Russian nuclear testing ground during the Cold War. A 50-megatonne hydrogen bomb, the largest ever, was exploded on the site in the 1960s. Novaya Zemlya became notorious in 1991 after an engineer at Murmansk publicly disclosed the amount of nuclear waste secretly dumped by the northern fleet and the ice-breaker fleet, Atomflot, in the various gulfs off the island and in the Kara Sea. It is now estimated that more radioactive liquid and solid waste has been dumped here than in all other oceans. Between 1965 and 1966 four submarine reactor compartments were scuttled in the Abrosimov Gulf in 20–40 metres of water and in 1967 three reactors from the ice-breaker ship *Lenin* were blasted to the bottom of the Sivolky Gulf. A barge with an underwater reactor was sunk in the Kara Sea in 1972. After an emergency in 1982, the submarine *K-27* with two fuel-laden reactors was dumped in the Stepovov Gulf. In 1988, a reactor was dumped in the Techeniya Gulf. Over the last thirty years between 11,000 and 17,000 canisters of solid radioactive waste have been dumped in the sea. It is rumoured that sailors cut holes in the canisters in order to sink them. The sailors thought that if the caesium was detected, it would be assumed that it had come from the nuclear testing on the island or from Sellafield. High levels of caesium were recorded in the sea water. It is estimated that from 1961 to 1990 a total of 165,000 cubic metres of liquid waste has been released into the Barents Sea to the west of the island, amounting to 2 million curies of radioactivity. This underwater graveyard is of obvious concern to the fishing fleets in the area, particularly to the Norwegians. Yet, studies conducted by both Western and Russian scientists reveal that there is little at present to worry about. The radioactive contamination has remained local, binding itself to the glutinous clay on the seabed. The radionuclides that do not penetrate the bed, such as strontium, become so diluted that they are virtually untraceable. 'The northern fleet thing is actually a big yawn,' says Charles Hollister, of the Woods Hole Oceanographic Institution in Massachussetts. 'It didn't surprise us one bit that nothing has happened much. What is of more concern in the future is where the Russians will put the rest of the waste. A lot of oceanographers feel that they should simply drag those submarines out into the deep water and let them go. Anything else would be expensive and highly dangerous.'

8. *Fear of Frying*

The Phobia Society in Manchester had a patient with an obsessive compulsive disorder, described by the society's director as 'an irrational fear of radiation'. Having read that fallout from the Chernobyl accident had reached Wales, the patient became so terrified of being contaminated that she changed her doctor because he was Welsh.

A day after the accident at the Chernobyl power station, word got out that party officials in Kiev were telling their families to leave the area immediately because of the serious radiation release. Within hours there was blind panic at Kiev Station, as thousands of people fought their way on to trains in a mass exodus.

It is a paradox that in 1980, the year after the nuclear accident at Three Mile Island, there were 50,000 deaths and 500,000 injuries on American roads. Yet, 80 per cent of US citizens continue to drive without seat belts. At the same time, an opinion poll revealed that 50 per cent of the population was afraid of nuclear power, even though not a single death had been attributed to the near meltdown at Three Mile Island. Psychiatrists and experts in phobic thinking explain this paradox in a number of ways. To begin with, it is important whether a person feels that he or she is in control of a particular risk. Does the risk-taker view the risk as voluntary or involuntary? Is the risk one large catastrophic event, as in an aeroplane crash, or a series of small individual events, like lung cancer? Another factor is whether the risk is familiar to the individual, like driving a car, or unfamiliar, as in an earthquake? Is the risk considered to be necessary or unnecessary? When these principles are applied to nuclear power, it is little wonder that the fear is so acute.

'Radiophobia', or 'nuclear phobia', as it has been defined, is regarded by the nuclear establishment as irrational, misinformed and the root cause of actual physical illness. Dr Leonid Ilyin, of the Institute of Biophysics in Moscow, is a strong believer in the destructive qualities of radiophobia. He claims that the stress, upset and disorientation caused by relocating a million people from the vicinity of Chernobyl increased their fear of radiation, which has resulted in a general deterioration of health and, in some cases, death. The popular diagnosis for this deterioration made by the media and the environmentalists is, he says, radiation poisoning, which increases the fear further still.

'The part of the brain that causes panic is the *locus ceruleus.* It is the principal neurotransmitter and is the part of the brain that detects novelty,' says Dr Robert L. DuPont, a psychiatrist specializing in fear. 'It also detects excitement and fear. If it is stimulated over and over again it stops firing, so in order to conquer a fear you must keep doing whatever it is you are scared of. The difficulty, however, with the fear of radiation is that it is not like an ordinary phobia, you cannot tackle it in the same way as you would, say, a fear of flying or a fear of elevators, because it is not an "all or nothing" experience. Once the door of an elevator or plane closes, that's it; you're in and you can't get off until it stops. The other problem with radiophobia is that people generally don't want to cure themselves of the fear. Instead, they want to cure the world of radiation.'

Our concern about radiation is partly fuelled by a fear of the unknown. Radiation cannot be sensed by the body. It defies our built-in safety mechanisms: you cannot taste it (although some people claim to have identified a faint metallic taste in the presence of high levels of radiation), smell it, feel it, see it or hear it. Left to our own devices, therefore, we are helpless and are forced to rely upon government agencies or industry to notify us of radiation leaks or accidents. However, their candour cannot always be totally relied upon. It was not until 1995 that the true extent of the contamination from an explosion in 1977 at Dounreay nuclear power station in Scotland was disclosed. We now know that the US Government was fully aware of the Kyshtym explosion near Chelyabinsk in 1957, but remained silent, fearing that knowledge of the accident might damage their own nuclear plans.

At high levels, radiation passes through the body as a wave of energy that leaves a trail of destruction in its path. It destroys the central nervous system and the tissue structure. It affects the gastro-intestinal tracts and dramatically reduces the red blood cell and white blood cell counts. In some cases it can kill in a matter of hours. An almost immediate effect on the body is the loss of control over bodily functions. It causes diarrhoea, vomiting, blurred vision, loss of balance and, after a time, loss of hair, darkening of the skin and wrinkles. Many of the young engineers who were exposed during the accident at Chernobyl looked fifty or sixty years old when they died. Radiation at lower levels, however, has a much more insidious effect. Few, if any, definable symptoms show up immediately, but years later it can cause a plethora of cancers. In spite of the fact that we know more about radiation than we do about any other known carcinogen, it is still a matter of angry debate as to exactly how low radiation levels must be before the deleterious effects become undetectable. 'Radiation,' says Dr DuPont, 'is like rabies or AIDS. It is particularly prone to fear because there is this long lag period of uncertainty before the symptoms are realized. A great deal of fear is to do with uncertainty. If there is a slither of uncertainty, the fearful mind is set off, because the only acceptable risk to a fearful mind is zero risk, and science does not talk about zero risk.' A risk, unlike smoking, crossing the road or driving a car, that is unacceptable because it is involuntary, beyond our control. This in turn increases fears that the government and the industry are lying to us in a grand conspiracy; the result being, whether consciously or unconsciously, that pro-nuclear statements are 'big business' statements and anti-nuclear statements are 'public interest' statements.

Of course, nuclear power is associated primarily with the atom bomb. Hiroshima and the image of a mushroom cloud have defined the twentieth century. In other ways, too, the dropping of the A-bombs on Japan acted as a prelude to the nuclear era. This was already obvious in the mid-1950s when President Eisenhower delivered his 'Atoms for Peace' speech to the United Nations in 1953. The idea was to encourage the public to focus on the benefits of the atoms, rather than dwell on its destructive implications. In those optimistic times the dream of nuclear energy was regarded by many as the ultimate peace dividend. The optimism soon failed, however, and more recently its negative image has been felt deeply by the nuclear industry, whose bureaucrats have worked tirelessly but in vain to win over public attitudes. The Visitors' Centre is now an integral part of the nuclear power station in the UK; few, if any, coal burning or hydroelectric power stations feel the need for this kind of public relations exercise. Nuclear Electric operates what it calls an 'open-door policy' at their power stations. Three thousand people visited Nuclear Electric's visitors' centres between 1994 and 1995, and the numbers are increasing. 'We are moving into tourism in a big way,' a Nuclear Electric representative told me. Their visitors' centres, besides the usual slick audio-visual presentation, include nature trails, talks, a classroom for children and the opportunity of doing joint projects with the nuclear engineers. The thousand-acre site at Sizewell, on the Suffolk coast, has its own English Nature and English Heritage officers to take you on a tour of the surrounding countryside. But whilst thousands of tourists happily wander around a nuclear power station for the day, there would be little chance of persuading them to live near one. In fact, in emphasizing the safety aspects of nuclear power, the industry very effectively highlights the dangers as well. The fearful mind asks, 'If it's so safe, why do they have to take so many safety precautions?' Nuclear power is perceived as being inherently dangerous. Much of the fear of radiation is irrational and misinformed, of that there is little doubt. Fear, the psychiatrist says, is to do with feelings not statistics.

'I chose this place because it isn't threatening,' says Dr Robert L. DuPont, referring to his office in Rockville, a business suburb outside Washington DC. 'There are no elevators and you can see your car parked right outside, so you know you can escape quickly if you have to.' He was not talking about his own personal needs, but about his clients'. Robert DuPont, a Harvard-trained psychiatrist, is president of the Institute for Behavior and Health, which specializes in phobias. 'I get a lot of people who are afraid of going into elevators or supermarkets, people who are afraid of going over bridges or scared of snakes. I get all types.' He described his office as cosy, but that would depend upon your phobia. If you were phobic about large house plants, fluffy toys or thick-pile carpeting, Dr DuPont's office would be a no-go zone.

On 28 March 1979 at four o'clock in the morning, nuclear engineers at Three Mile Island in Pennsylvania made a disastrous mistake. They accidentally shut off the emergency water coolant to the second reactor's core and very soon the fuel elements began to overheat. A state of emergency was called on the island and mass panic ensued in the neighbouring towns of Middletown and Harrisburg as the residents evacuated the area. In fact, a core meltdown was narrowly avoided, and forty-eight hours later the situation was again under control. But the reactor was irreparably damaged – so was the American public's confidence in nuclear power. The years following the accident have seen no new orders for nuclear power stations in the United States, as well as a large number of cancellations.

'Three Mile Island meant nothing to me really,' says Robert DuPont, 'but it was my introduction to *nuclear phobia.*' At the time of the accident, DuPont was a well-known figure in the field of drug awareness. He was head of the District of Columbia's narcotics treatment agency and had been director of the National Institute on Drug Abuse during the Nixon and Ford administrations and part of the Carter Administration. In 1979 he regularly appeared on the television show *Good Morning America.* 'My work,' he says, 'has always been controversial and I've enjoyed the controversy. I remember going on a black radio talk show in Washington DC to be interviewed about methadone treatment as a way of getting people off heroin. There was a feeling at the time that the Government was deliberately selling drugs to black people to keep them tranquillized. A spokesperson for the black community in DC called in and advocated that all black people should get together to kill me, because I was destroying their community.' In 1984 Robert DuPont was once again at the centre of a heated public and professional debate concerning his work on the fear of nuclear power. He denies that the debate damaged his career, but adds, 'I was never a big player, but after that I was certainly off the field. It was sort of like being asked: When did you stop beating your wife? It was a no-win situation.'

In 1980 Dr DuPont, then president of the Phobia Society of America, coined the phrase 'nuclear phobia'. 'I was contacted by a journalist who was writing a book about problems in media coverage. He asked me if I would watch a network TV coverage of nuclear power since the late 1940s. For two solid days I sat and watched thirteen hours of television. Sixty per cent of it was about the accident at Three Mile Island. My impression was that the dominant theme running through almost all the coverage was fear. The media dealt with the accident not as a "what is" situation, but as a "what if" situation. The questions constantly asked were: What would have happened if it had exploded? What if there had been a huge radiation leak? Yet we knew that what had happened was that the problem had been contained and no one had died or been severely injured.' Dr DuPont's subsequent paper, *Nuclear Phobia – Phobic Thinking about Nuclear Power*, was enough to get him hooked on the subject. Over the next few years he became the expert on the subject and travelled the country giving after-dinner speeches to scientific associations, while continuing his research by visiting power stations and talking to their local communities. 'I would simply talk to people who lived within site of a reactor about what they were afraid of. Rationality, I discovered, had nothing to do with their fears. I remember going to Three Mile Island, my wife was frightened for my health. I was a little anxious myself. What I found, however, was the most routine, mundane place. There was this woman selling T-shirts in Middletown and she could see the workers wandering around the plant from her house. She wondered how it was that they could continue working there when the rest of the world was scared to death. You get this incredible disconnect with nuclear power, where the workforce have no fear at all and the outside world is terrified.'

In 1984 DuPont put in a grant proposal to the Department of Energy (DoE), under the heading, 'The Psychology of the Phobic Fear of Nuclear Energy'. 'The DoE research department gave me US$85,000 – peanuts! But the guy who ran the department was interested in the subject.' DuPont is quick to stress that they had no input into his research, nor was there any hidden agenda. However, news of the grant ignited a reaction from the media and in the psychiatric world that DuPont finds baffling to this day. Howard Kurtz of the *Washington Post* picked up the story and it became front-page news on 30 October. The article began: 'The Department of Energy, on the theory that people who oppose nuclear power may suffer from an irrational phobia, is paying a Rockville psychiatrist US$85,000 to find out if their fears can be overcome.'

'I could not believe it,' recalls DuPont. 'What the story was saying was that I was labelling people who were afraid of radiation as mentally ill, which was an abuse of my medical role. In other words, I was doing what Hitler had done . . . a totalitarian, horrible thing.' That morning CBS, NBC, ABC and *The New York Times* queued outside DuPont's 'cosy' office, waiting their turn to get their pound of flesh. 'I was completely confused by the whole thing. I had no idea what I had said or done wrong. What was worse was that my friends all assumed that I was in the wrong. I had somehow ceased to be a person and had become a symbol, which was not very pleasant. I felt so alone.' He replied to Howard Kurtz's piece in a *Washington Post* editorial. 'Howard Kurtz,' he wrote, 'created a halloween nightmare for me, as media representatives descended on me, howling objections to my waste of taxpayers' money and my attempts to use psychiatry to stifle political debate. Mr Kurtz wrote that I had "convinced the Government that fear of a nuclear accident could be a psychiatric disorder". I have repeatedly emphasized that those who fear nuclear power are not suffering from a diagnosable disorder, nor are they mentally ill.' The story ran for two weeks. 'Dr DuPont wouldn't know a nuclear plant from a McDonald's,' ran an editorial in the *Washington Post*, 'DuPont is, like the DoE, a booster of nuclear power.' 'What is not excusable is the mind set that would prompt a federal agency to elevate to the level of scientific inquiry the notion that its opponents are the victims of a psychiatric disorder' . . . 'There is nothing pathological about the general public's fears.' *USA Today* observed incredulously, 'Mr Phobia says it's far more dangerous to ride a bicycle around the block once than it is to live next to a nuclear power plant.'

At the annual meeting of the Phobia Society, Dr DuPont met with little sympathy among his colleagues. 'As a bunch, we psychiatrists are fairly liberal and mostly anti-nuclear. Suddenly, I was viewed as pro-nuclear and right wing, and therefore I had betrayed the underlying ethos of the profession.' Dr Manuel Zane, one of DuPont's mentors and someone he revered greatly, criticized him in the *Psychiatric News* for being naïve. Zane wrote that DuPont used the word 'phobic', while at the same time protesting that he does not regard 'radiophobes' as being mentally unstable. 'To them and to me (though not to him) the term phobic implies a pathological psychiatric basis for their fear.' He went on to criticize DuPont for assuming that nuclear power was safe, despite the fact that almost half the country, including a great number of respected scientists, thought otherwise. 'That really hurt,' says DuPont of an article by Dr David Musto, a professor of psychiatry. 'I've known David, pretty well, for 25 years, but he made me look like a buffoon.'

'The Good Doctor has prepared a brochure so that he may determine what best allays their fears. He will then turn over the perfected brochure to the DoE,' wrote Musto. 'Dr DuPont has long been impressed by the remarkable safety of nuclear energy; he is so struck by the contrast between the industry's excellent safety record and Americans' fear of nuclear technology that in 1980 he coined the term "nuclear phobia" to describe the irrational fear of atomic power . . . Dr DuPont called the fear underlying the energy policy debate a "curious type of fear."'

Today nuclear phobia is a fear that Dr Robert DuPont hopes to forget. In any case, he is unwilling to continue his research. He is, however, anxious to clear his name. 'My job,' he says, 'is to put fear into perspective. I spend my life getting people to ride in elevators, inducing them to walk into shopping malls or on to aeroplanes. All I wanted was to get people to go around nuclear plants. I still believe that the key to this thing is familiarity; the more that people are exposed to nuclear power the less political it will become and therefore the fear will subside.' As we were parting he told me of a woman he had met while visiting the Diablo Canyon nuclear power plant in California. 'This woman advised me to stop my research because it was too big and powerful for me. I interpreted her warning as another example of fear of a government and industry conspiracy. Maybe I should have listened to her.'

Edward Teller is widely regarded as the model for Dr Strangelove. But when I met Teller in Edinburgh at the end of 1994, he struck me as being very far removed from the obsessive power-crazed scientist so vividly portrayed on screen by Peter Sellers in the Stanley Kubrick film. Teller was visiting a fellow-Hungarian at the time, one of his disciples, who lived on the outskirts of the city. Nevertheless, he granted me an interview over lunch and sat for an hour contemplating a plate of sandwiches as he parried my questions with barely disguised ill will. On the way home I couldn't help wondering if Kubrick had ever heard of Teller's Soviet counterpart. Now director of the experimental research station at the secret Russian atomic plant Chelyabinsk 65, Dr Gennady N. Romanov had only just started his career when Teller was experimenting with the hydrogen bomb. The mood of the Cold War is encapsulated in his personality. Like Teller, Romanov is a man possessed by a firm, unshakeable belief in the ability of science to dominate nature. In this respect he is typical of a generation of scientists. The 1950s were, and continue to be, an atomic dream to Teller and Romanov. During the 1950s and 1960s military leaders and industrialists dreamed up a multitude of bizarre applications for the power of the atom. The Sahara desert was to become a forest; there were to be atomic planes, trains and helicopters which would go faster and operate more efficiently than could be imagined. In addition, nuclear power was to be deployed not only to melt icebergs, to flatten mountains and to control the weather, but also to provide energy too cheap to meter. It was suddenly patriotic and glamorous to be a nuclear physicist. Scientists were celebrated and cherished. They appeared on the front cover of *Time* magazine, and were rewarded with new laboratories, political support and government funding. In a symptomatic flight of fancy, which now reads like something out of H. G. Wells's *The Shape of Things to Come*, Glenn Seaborg, chairman of the Atomic Energy Commission, once said of a planned nuclear rocket, 'What we're attempting to make is a flyable, compact reactor, not much bigger than an office desk, that will produce the power of the Hoover Dam, from a cold start, in a matter of minutes.'

None of these dreams came to anything, but their abandoned prototypes lie scattered in places like Hanford and Nevada and in the shipyards of Murmansk. Seven-and-a-half-metre-tall jet engines that held nuclear reactors lie rusting in Idaho. Two engines weighing approximately 250 tonnes each would power a nuclear bomber carrying a full load of atomic weapons for weeks, possibly months, without refuelling. In Russia the few remaining atomic ice-breaker ships are now used for nothing more than the occasional tourist trip to the Arctic.

I met Dr Romanov in a government building on the main street in Chelyabinsk. He had come from Chelyabinsk 65, the nuclear weapons plant 115 km north of the city, to speak to me. A serious man in his late fifties, immaculately dressed in a dark suit and tie, with his hair cut short and scraped back, he was obviously regarded by his political masters as a reliable frontman to put across the official version of the secret atomic plant. Like most Russians he is a chain-smoker. In 1958, while still only a young man, Romanov was given the job of setting up the experimental research station at Chelyabinsk 65 to study the effects of radiation on the environment, or 'radioecology' as it is now called. Romanov commands the research station with a strict military discipline. Like many who have dedicated a career to the nuclear weapons industry, he is on guard and defensive in the company of outsiders. As far as Romanov is concerned, one is either inside or outside. He told me a story about a Russian general who had run, efficiently, the secret city where the SS18 ballistic missile was manufactured. At the end of the Cold War a team of Westerners were granted access to the plant in order to assess the numbers and the condition of the weapons. The event proved to be too much for the general. Suddenly and dramatically his whole existence had been brought into question. After the team left he put a gun to his head and shot himself.

Romanov's research centre was set up in response to the Kyshtym explosion. In September 1957 an underground high-level waste tank overheated and exploded as a result of a leak in the cooling system.

What is known as the East Urals Radioactive Tract contaminated 23,000 square kilometres of land to the east of the nuclear plant, leaving a population of 272,000 people exposed. Just over 10,000 people were resettled in all. The most contaminated land is that nearest to the site of the explosion. Romanov's research centre occupies 16,000 hectares of that land. He likes to refer to it as his 'national park', because far from the brittle, dead environment one might expect, it is flourishing 'like the garden of Eden', he boasts.

Romanov started the research by discovering how radionuclides (mainly strontium-90) behave in the environment. He claims that within three years of the explosion they had begun to recover the polluted territory, using nothing more than a plough. The plough was specially designed to remove the contaminated top soil and bury it to a depth of 75 cm. By constantly turning the soil in this manner, the levels of strontium-90 were reduced tenfold.

The second phase of research was to study the effects on farm animals grazing on contaminated pastures. Several experimental farms were set up. Romanov discovered a significant reduction in the uptake of strontium-90 as it passes down the food chain. According to his research, a plant accepts in equal measures non-radioactive calcium and radioactive strontium-90 from contaminated soil, whereas a cow's stomach is able to discriminate between the two and absorbs the calcium in much greater quantities than the strontium-90. 'We have standards for acceptable levels of strontium-90 in the food that we eat from this area, and the beef from our farms is within those standards,' he told me. 'If the population only ate food from the contaminated area, then they would get an average dose of only 3 rems a year. That would be within the acceptable levels.' Romanov realizes that much of what he says is controversial and contradicts what the public expect to hear. When his work was declassified, he was criticized by Western scientists for trying to conceal the effects that radiation had on the environment. 'Everyone wanted me to talk about mutations,' he says, shrugging his shoulders. 'Public faith in science and scientists was ruined by Chernobyl. No one trusts us any longer. The politicians are at fault for blaming the scientists for the accident. In many ways the secrecy that existed in my day was a good thing because it did not exercise public opinion.'

Romanov is an oddly and unexpectedly charming character. He can make his rather macabre assertions sound utterly reasonable. As we talked, he was obviously enjoying himself. He likes an audience and savours the extreme reverence paid to him by his colleagues. His professional status has become its own reward – it's all that's left. Because, although Russian nuclear scientists were among the most privileged members of the *nomenklatura*, enjoying large houses and chauffeur-driven cars (Yuli Khariton, the father of the Russian A-bomb, had his own railway carriage) and holidaying in dachas on the Black Sea, they never received the publicity and renown extended to their American counterparts, such as Teller and Oppenheimer.

Romanov helped himself to coffee and cake off a trolley that had been pushed nervously into the room. 'The problem is that atomic energy and all civil uses of the atom have become associated with the bomb,' he said, between mouthfuls. 'People immediately picture death and genetic deformities. They must understand that it is not all frightening. There are safe levels of radiation.' In Russia many of the nuclear establishment believe in *hormesis*, the idea that a little radiation is good for you. In the West the theory is known as 'adaptive response' and is hardly paid any lip-service. It argues that small levels of unnatural radiation can enhance the life span of cells. The theory behind *hormesis* is that a cell can repair minor damage to itself, damage that might be caused naturally by background radiation, so if that cell is provoked into repairing itself on a regular basis by giving it small doses of unnatural radiation, the cell will remain healthy longer.

Romanov's views on nuclear power also toe the establishment line. He regards it as inevitable and necessary; with the future of fossil fuels looking so finite, nuclear power is the only power source at present that has any lasting future. 'We have to convince people that, yes, there is a risk from nuclear power, of course there is. We cannot help irradiating people and polluting the environment, but that risk is acceptable. We must teach the people to put that risk into context with all the other risks, like smoking, driving or the risk from chemical pollutants. People will eventually understand and accept this, but probably not in my lifetime,' he laughs. He uses Chernobyl as an example. 'Those people who were relocated as a result of the accident received a dose on average of about 12 rems. This is not dangerous, what is much more dangerous is to relocate them. Relocating these people affects their livelihood and pride. They get treated as lepers by the communities they move to and they suffer from heart disease, arthritis and a host of other problems. They blame everything on radiation. But their illnesses are a result of the life change and their own radiophobia.'

The last time an American psychiatrist involved in a government-funded study spoke publicly about radiophobia, he was castigated by the media and branded a lunatic. In Russia people talk about radiophobia all the time, blaming it for a myriad of health problems and complaints. One of Romanov's favourite stories was of the two Russian officers in Nagasaki who, only days after the end of the Second World War, inspected ground zero, the epicentre of the explosion, in order to witness the damage inflicted by a single bomb. One was paralytic with vodka, the other was sober. The sober man died only a few years after the assignment. The drunk man is still alive. To Romanov this proved that radiation is really nothing to worry about. In fact, he goes a step further – he admires radiation.

Romanov went on to tell me about his favourite part of the research. This involved firing particles of radiation at selected parts of the land using a mobile radiation emitter, a device that looks like a cannon mounted on a trailer. The emitter enabled Romanov to identify what doses affected which parts of the environment. Travelling around the research centre's land on the back of a tractor towing this home-made contraption, Romanov was able to irradiate different species of trees, grass, plants, roots and seeds. I couldn't help thinking of Robert Duvall as the mad colonel in *Apocalypse Now*, who loves the smell of napalm in the morning. Here was Romanov's *pièce de résistance*; zapping nature beats ploughing top soil, any day. 'In order to kill a woodmouse, we had to give it a total dose of about 5,800 roentgens,' he grinned. 'To kill a birch tree, we would give it 20,000 roentgens over two or three days, and to kill a pine tree, which has a resistance similar to humans, we have to give it 2,000 roentgens. These doses are of course huge and would only occur in the event of a very serious accident,' he was quick to add. Having identified individual lethal doses for each living organism, he began to irradiate a hectare of land at a time, using a dose that is estimated to kill everything. To Romanov's surprise, some mice, seeds and roots survived, and in less than eight years the irradiated hectare had repaired itself in the same way that the control groups had. A population of woodmice had been irradiated with a dose that was two to three times their individual lethal dose. A sufficient number of mice had survived the assault to continue reproducing. Now, seventy-two generations later, Romanov says that although there are changes to their bone marrow, there do not appear to be any visible mutations in their offspring, nor have they adapted their behaviour.

At the end of our talk Romanov offered to give me a lift to my next appointment in his chauffeur-driven official car. On the journey we got talking about America, which he had recently visited for the first time in order to attend a world radioecology conference. He disliked the place. 'They are wasting billions of dollars on cleaning up their nuclear sites. Do they think that pollution to the environment is avoidable?' His other complaint about America was the quality of the food, but it seemed odd coming from a man whose home cooking is laced with strontium-90.

Every Sunday evening, between 6.30 p.m. and 10.30 p.m. Princeton 103.3 FM plays folk music. John Weingart, disc jockey for the last twenty years, has a loyal following, but the radio station is more of a hobby than a way of life. Weingart has one of the world's most unpopular jobs – the executive director of New Jersey's Radioactive Waste Disposal Facility Siting Board. He wishes he could think of a grabby acronym or a catchy slogan for the job. 'The environmental movement dream up such good ones,' he says wistfully. Recently, at his 25-year college reunion, he was asked by his contemporaries what he did for a living – and he told them. 'There would be a long silence,' he recalls, 'and they'd look at me as though I'd sold my soul to the devil.' In a cartoon pinned to the wall of his office a boffin explains to his assistant that he is aiming for 'a series of commercials in which

I, through the sheer force of my personality, make radioactive waste sexy!' Weingart's job is to explain to the people of New Jersey the benefits of having a radioactive waste dump situated in their backyard. He considers himself to be a liberal sort of person. Most of his friends tend to be sceptical about nuclear power. He admits that he felt the same way before he started the job, but has now come to realize the weakness of the anti-nuclear case. 'It is all about education,' he adds. 'My friends listen to my views now and even though they still do not consider it as an acceptable risk, they think about it a little longer than they used to.'

In 1980 the Federal Government passed the Radioactive Waste Policy Act, handing over responsibility of finding waste sites to individual states. Each state had the option of forming a compact with other states. These compacts were exempt from the Administrative Procedures Act and therefore were not obliged to hold hearings or public enquiries. New Jersey joined the north-east compact with eleven other states, but negotiations broke down after they failed to agree which state would ultimately accept the waste. Since then New Jersey has formed a loose coalition with Connecticut, but is more or less going it alone. The siting board was set up in 1987. 'The traditional approach in this game was to look at a map of your area and pick ten suitable locations,' says Weingart. 'Teams would then go out to those locations and explain to the communities there that they had been chosen as a possible site for nuclear waste, and do they have any questions? At which point the residents' immediate response was to call their lawyers.' At the beginning of 1995, however, the siting board tried a different tack, one that is beginning to be favoured by others: the 'voluntary' siting policy. The theory is that a more *laissez-faire* approach, involving public debate and education, in addition to hard cash, will pay dividends. Weingart started applying the 'voluntary system' by writing to every Rotary Club in the state, advertising himself and his chosen topic of conversation. 'There's some poor sucker at the Rotary Clubs who has to find fifty speakers a year. They'll take virtually anyone, so long as he's a warm body!' Until now he has given his fifteen-minute talk to 567 municipalities in New Jersey. He admits to it being frustrating. The most frequently asked question occurs when people shake his hand and enquire whether it will make them glow in the dark.

'There are two principal questions raised over the siting of waste: Is it safe and is it necessary?' says John Kelly of the consultancy JK Associates. 'There's a trade off between the two. People will usually accommodate the first, if the answer to the second question is yes. With nuclear waste, that has to be a big yes.' JK Associates specialize in the fear of nuclear technology. Their clients include the Department of Energy, Congress and the nuclear power utilities. 'Much of our current work,' says Kelly, 'is concerned with the moral and ethical arguments that have been forgotten. The Nuclear Waste Policy Act in America talks about our responsibility to deal with the waste now, rather than leave these problems to future generations. We must honour that responsibility.' Kelly argues that the issue of whether a site is necessary, is being forgotten in the debate. 'Too many people concentrate the debate on the safety aspects of the waste site. The only acceptable form of safety becomes zero risk, which is ridiculous. There's no such thing as zero risk. But when it comes to radioactivity, it is the only type of risk that's tolerable. What the public do not think about is the alternative: What is the risk from ten or twenty smaller sites as opposed to one large site?'

John Weingart is a tall, slim man in his mid-forties. Unlike many people in the nuclear industry, he is neither arrogant nor smug. A slight awkwardness gives him the air of vulnerability. He admits to a feeling of isolation. His talk starts by emphasizing that 60 per cent of New Jersey's energy is created by nuclear power stations. 'As a result,' he tells the audience, 'we have to face up to the responsibility of dealing with the 570–850 cubic metres of waste generated each year.' He explains that this amount would be equivalent to eighteen double garages. Ninety-three per cent of its radioactivity from that waste comes from New Jersey's four power plants. The rest comes from hospitals, pharmaceutical companies, research facilities. In return for housing the waste, Weingart offers communities US$2 million a year for the next fifty years. He guarantees that open spaces will be left undeveloped and promises to improve the infrastructure of the community. He is also willing to protect the value of the property in the area. Other than that, it's up to the community to negotiate. If the people want a new school, sports centre or old-people's home, John Weingart is willing to talk about it. All the siting board requires in exchange is an 'acceptable' 50-hectare site in a relatively isolated spot. This narrows the choice significantly as New Jersey is the most populated state in the US. The site should have good access, meet the right environmental criteria and, most importantly, there must be a high level of acceptance to the plans in the community. 'My way is non-threatening. I tell people that they are going to get a good deal, but that they are free to ignore the offer if they wish.'

We watched a video the morning I visited Weingart in his office. He had just got back from his holiday and had missed the news report. The report told the story of the small town of Hamburg, New Jersey. It started with images of children playing at Hamburg's main attraction, the Gingerbread Castle. The next shots were of trees, flowers, wildlife and the reporter standing waist-deep in a wheat field. 'Would you want this,' the reporter gestured at the pastoral scene behind him, 'to become a nuclear waste dump?' The item went on to explain how a local farmer had contacted John Weingart to talk about the possibility of a waste site on his land, but before the talks could begin, the inhabitants of Hamburg and the neighbouring town had organized themselves in solid opposition. 'That's what we're up against,' said Weingart disappointedly, as he switched the video off. 'The issue is so political that no one, particularly those with enough power and money to influence public attitudes, can afford to even discuss the subject.'

'There is a distinction made between good and bad radiation,' he continued. 'Any radioactive waste from a nuclear power station is bad radiation, whereas any radioactive waste from a hospital or medical research institutes is good radiation. At the moment we couldn't sell a waste site if all it contained was newspapers left by nuclear workers at the gates of a nuclear power plant.' One problem is a lack of credibility. The siting board is primarily funded by the New Jersey nuclear power industry. The result is that the public trust the environmentalists instead, and the more extreme their view, the more they believe them. 'Environmental groups turn up at meetings and start

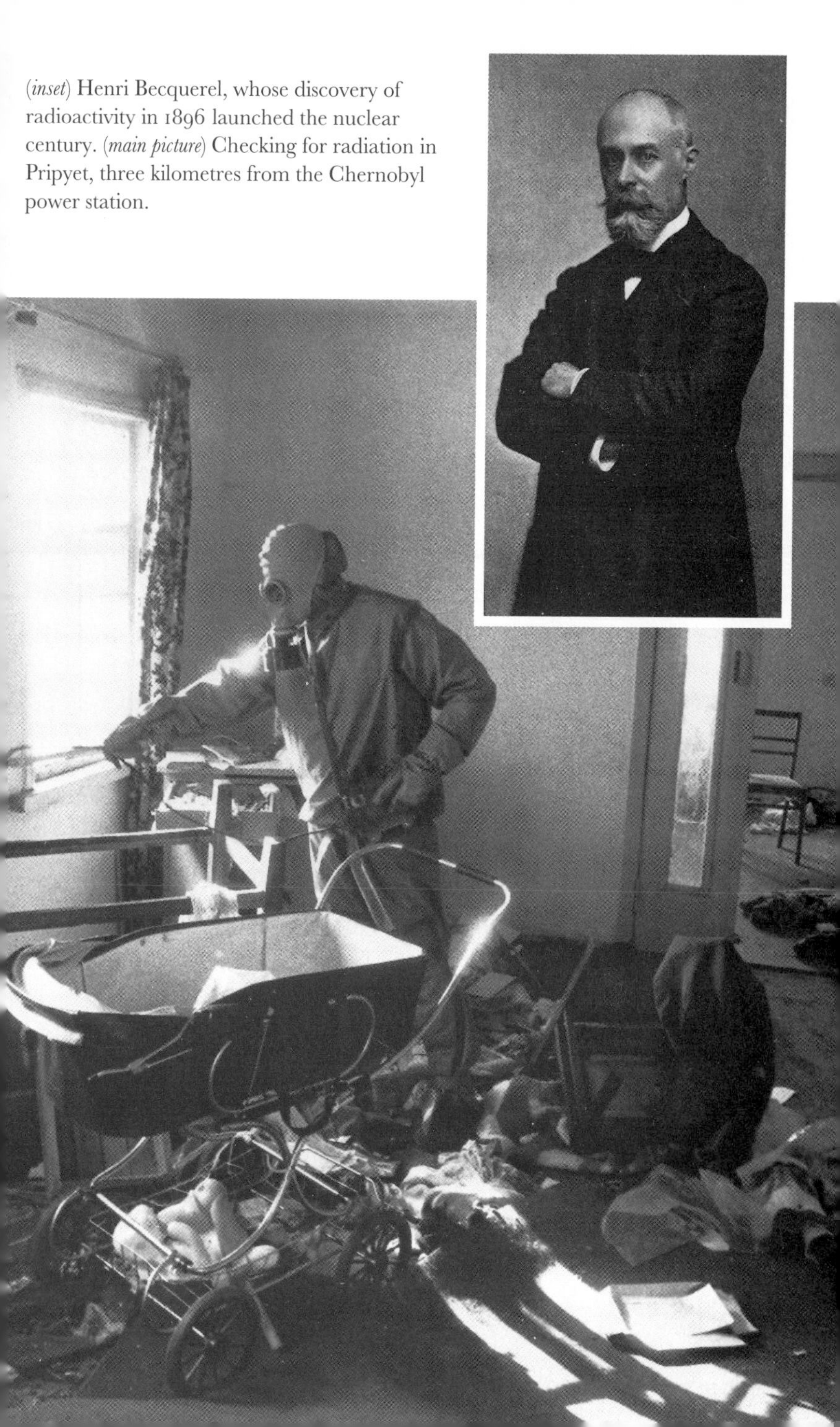

(*inset*) Henri Becquerel, whose discovery of radioactivity in 1896 launched the nuclear century. (*main picture*) Checking for radiation in Pripyet, three kilometres from the Chernobyl power station.

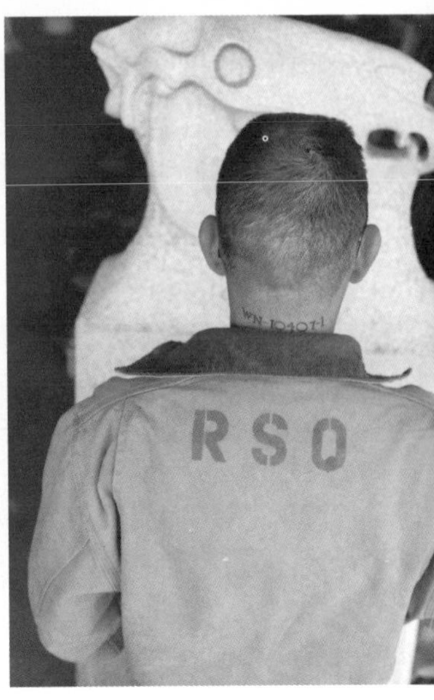

(*above left*) James L. Acord, nuclear sculptor, with his radioactivity licence number tatooed on the back of his neck (*above right*). Gary Lekvold (*below*), the Hanford whistleblower.

The elusive Wendell Chino (*above*), leader of the Mescalero Apache Indians, with members of his tribal council. (*below*) Rufina Laws, anti-nuclear protester and dreamer of dreams.

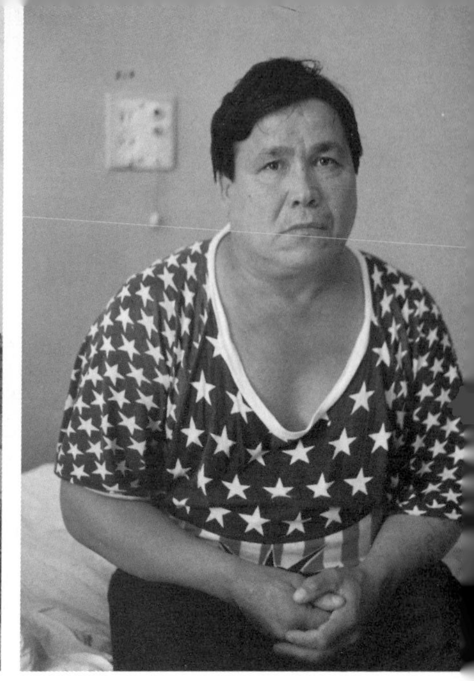

(*above left*) The Techa, the world's most radioactive river. (*above right*) A radiation victim from Muslimova. (*below*) Sharia Khmetova and fellow patient at the Urals Research Centre for Radiation.

(*Top*) Mira Kossenko, head of epidemiology at the Urals Research Centre, examining the health records of Muslimovans. (*middle*) An orphanage for radiation victims in Chelyabinsk, the 'most polluted spot on earth'. Edward Teller (*below*), inventor of the H-bomb, with the author in Edinburgh, September 1994.

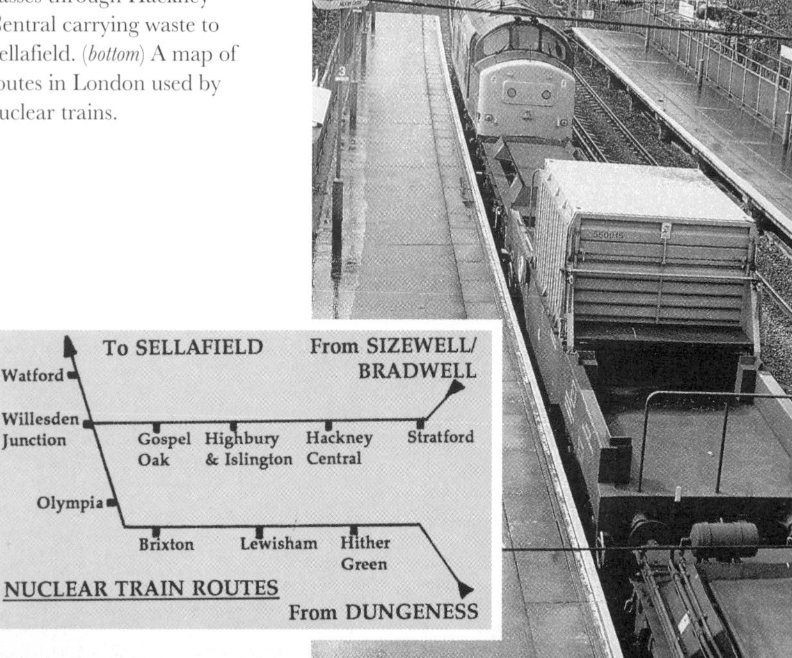

(*top*) Greenpeace protest against the *Akatsuki Maru*, the first shipment of plutonium to Japan. (*middle*) A nuclear train passes through Hackney Central carrying waste to Sellafield. (*bottom*) A map of routes in London used by nuclear trains.

Tom Johansen (*right*), the used car salesman who purchased a nuclear bomb kit from the United States government. (*below*) A mutant scorpion fly painted by Cornelia Hesse-Honegger. (*bottom*) A street in Schlema, east Germany, where uranium for the Soviet nuclear arsenal was mined.

'Not in my back yard': growing up in the shadow of Sellafield nuclear plant, west Cumbria.

telling the public rubbish like if there was a radiation release, everyone within an 80-kilometre radius would die in an hour or every child would get leukaemia.' Another problem is the exact definition of low-level radioactive waste. In the United States, low-level radioactive waste is classified by what it is not. It is not spent fuel rods or uranium mill tailings. But apart from that it can be anything. This creates a great deal of confusion and distrust. Weingart acknowledges that this is a problem, particularly as low-level waste can contain up to 100 nanocuries of plutonium per gramme. 'We define it as waste that has decayed to background radiation levels within 500 years. Which is how long ago Columbus discovered America. A point I am constantly reminded of,' he smiles.

Nevertheless, Weingart is optimistic – he is convinced that sooner or later a site will be found. I asked him if he would be happy for the site to be next door to his own home. He lives with his wife and children on a small farm in a rural part of New Jersey. He thought for a minute. 'I don't know,' he said, 'I just don't know.'

PART FOUR

9. *Ships, Trains and Wastemobiles*

At the end of the Cold War Greenpeace observed famously that the Soviet Union had contributed a vital link in the bizarre delivery of plutonium to the United States' nuclear arsenal. The allegation exposed the magical mystery tour that lay behind the international nuclear trade. It began with uranium from Canadian mines being shipped to Cap de la Hague in France, where it was processed into uranium hexafluoride. The 'hex', as it is known, was then sent to Riga in Latvia for enrichment. The enriched uranium was shipped back to Cap de la Hague and then to Seattle on the west coast of America, where it was turned into uranium oxide. The material was transported by train to New Jersey for conversion into fuel rods. The fuel rods were shipped across the Atlantic to West Germany, where they were used in reactors. The spent fuel from the German reactors was sent back to Cap de la Hague for reprocessing, where it was finally separated into depleted uranium, to be returned to Latvia for enrichment, and plutonium which was later shipped to the United States and used in nuclear warheads.

Nuclear materials are transported around the world on a daily basis. The International Atomic Energy Authority in Vienna documents approximately 10 million separate movements of nuclear material each year by road, rail, sea and air. As the Greenpeace story above illustrates, the international nuclear industry is constantly trading its wastes, reshaping its assets and using a variety of diverse skills, processes and reprocesses to sustain itself. A more recent example of the nuclear trade involved the Department of Energy's contractor at Hanford, Westinghouse. While decontaminating PUREX, the plutonium extraction plant, Westinghouse found a buyer for 900,000 litres of radioactive liquid waste. The waste, a contaminated nitric acid containing 7.5 tonnes of uranium, would have cost the Americans several million dollars to dispose of. British Nuclear Fuels Ltd (BNFL) at Sellafield, on the other hand, regarded the waste as nearly 10 per cent of their annual nitric acid consumption for the reprocessing plant at THORP. As part of the deal, BNFL agreed to extract the uranium and to ship it back to the US. With the rising costs of waste management, this sort of exchange is likely to increase in the future.

Transportation is a weak link in the nuclear process. Not only does it bring highly radioactive and toxic materials into close quarters with the general public, it also exposes the lethal cargo to a number of unpredictable risks, such as acts of terrorism, loading and unloading accidents, vandalism, collisions with other traffic, obstruction by nuclear activists and delays. It is also the one time that the nuclear industry really exposes itself; the knowledgeable and patient observer can guess with accuracy what a military site or nuclear plant is up to simply by watching the shipments in and out. As a result, public information concerning the transportation of radioactive materials is not readily available. It takes the work of environmentalists and campaigners, or the publicity from an accident or near accident, to bring the issue to light.

When the *Mont Louis*, a French cargo ship, collided with a Belgian cross-channel ferry in August 1984, it took four days for the Belgian Government to admit that the *Mont Louis* was carrying thirty canisters of uranium hexafluoride. The ship's register, according to one report, had the cargo written down as 'medical supplies'. Although the canisters were retrieved intact and there was no evidence found of a leak, the news of the accident had the nuclear industry running for cover. Articles followed that revealed that BNFL had been using Sealink passenger ferries to transport uranium and uranium hexafluoride to France and Holland. BNFL denied the report until a consignment note bearing the company's name was produced. Three days after the *Mont Louis* sank, the *Pacific Fisher*, a vessel owned by Pacific Nuclear Transport Limited (PNTL), was refused port services while going through the Panama Canal because it was carrying spent fuel from Japan to Sellafield for reprocessing. It turned out that the British nuclear fleet had been using the canal as a matter of routine.

Long-haul flights carrying plutonium were a popular mode of transport until the 1980s, when regulations were tightened. In 1980 an

American flight carrying plutonium to Japan was refused permission to ground and refuel in Alaska. Under the Nuclear Co-operation Agreement between the United States and Japan, air cargo of plutonium is not allowed into Japan unless it complies with rigid regulations. As a result, shipping is viewed as the only viable alternative.

In terms of public opinion, the transportation of nuclear waste on the high seas has caused more anger and condemnation than any other form of haulage. The case of the *Akatsuki Maru*, for example, and its 25,000-kilometre voyage to Japan, with 1.7 tonnes of vitrified plutonium on board, seemed to prove that the nuclear industry places big business before the environment. The *Akatsuki Maru* left the port of Cherbourg in November 1992, escorted by a new ship, *Shikishima*, armed with two 35-millimetre cannons, two machine-guns and two helicopters. The departure was held up by thousands of demonstrators at the port, and by the efforts of the Greenpeace vessel *Moby Dick* to block the ship's path. The only country not involved in the shipment to know the route that was to be taken was the United States, which had been asked to monitor the consignment by satellite. The ship made its way into the Atlantic and was chased by the Greenpeace vessel *Solo* until it was rammed by the *Shikishima*. The protests continued as the ship sailed around the Cape of Good Hope, across the Indian Ocean, past the southern tip of Australia and north, through the islands of the South Pacific. The Asian Pacific Forum called for Japan to halt the shipment. The governments of Portugal, South Africa, the Philippines, Malaysia, Chile, Brazil, Argentina and Indonesia requested that the ship stay out of their territorial waters. Australia and New Zealand also made official complaints. On arrival the *Akatsuki Maru* was greeted with further protest. Sixteen helicopters and sixty-nine vessels had to escort the cargo into the harbour. The entire exercise was estimated to have cost the Japanese Government 6.3 billion yen (£40 million).

In many ways, the widespread condemnation of the *Akatsuki Maru* shipment paid tribute to the very nature of plutonium, which is thought to be the most dangerous substance on earth. The possibility of an accident occurring that would lead to a release of plutonium into the world's oceans was (and still is) viewed as a substantial risk. Permissible levels for plutonium in the body are lower than any other radioactive element. It emits alpha particles that cannot penetrate paper but, if ingested, concentrate on the bone, irradiating the surrounding tissue, causing damage to DNA, as well as cancer. Plutonium-239 has a half-life of 24,400 years. The element was created artificially in December 1940 by Glenn Seaborg and two of his colleagues at Berkeley University in California. It was called 'Pluto', not after the god of the underworld, however appropriate that may seem, but because it came after Uranus and Neptune in the sequence of elements. Created at a time of war, its fissionability was immediately viewed as a useful component for the Manhattan Project – and ever since, its primary function has been to make nuclear weapons. It is this fact that makes the element so politically sensitive. The issue of proliferation, together with its longevity and toxicity, makes plutonium an overabundant and dangerous resource. There is estimated to be 1,200 tonnes of plutonium in the world: about 200 tonnes have been used for making bombs and the rest is a by-product of the nuclear industry. The UK has about 100 tonnes of 'civil' plutonium. The size of the military stockpile is, of course, a secret. Things have changed considerably since the fateful day when a small metal container carrying plutonium left Hanford for Los Alamos, to be used in the first atomic bomb explosion in the New Mexico desert, and later in the bomb dropped on Nagasaki. Ironically, fifty years on, the Japanese Government has just confirmed its commitment to the fast breeder reactor, and to plutonium as a major part of its future energy policy. Such a commitment now makes Japan a world leader in the sale and procurement of the world's deadliest material. As a result, many more global shipments are planned.

Following the *Akatsuki Maru*'s unhappy journey, the British ship *Pacific Pintail* fared no better upon returning the first batch of nuclear waste to Japan. The 14 tonnes of plutonium waste on board were being delivered to Japan after reprocessing at Cap de la Hague in France. The cargo contained 13 million curies of radioactivity. In February 1995, after a two-month journey during which it had been followed by Greenpeace, the *Pacific Pintail* negotiated its way towards the port of Mutsu Ogawara in the province of Aomori in northern Japan. At the last minute, upon the orders of Governor Morio Kimura, the ship was refused permission to enter the port. The governor, unhappy with the lack of a permanent storage plan, feared that the province of Aomori would become a nuclear-waste dumping ground. Ten years earlier, the Aomori Government had agreed with the Japanese Government that it could 'temporarily' store nuclear waste, for a maximum of fifty years, in underground storage pits at a nuclear-fuel reprocessing facility that was still under construction at Rokkasho. In return the Government agreed to subsidize the fishing and farming region. To date 40 billion yen (£260 million) has been given to the area. Morio Kimura was concerned that the Government would not honour the agreement. Another worry was that in December 1994 the province was struck by an earthquake that left much of the area destroyed. The tremor cracked roads and damaged a quayside at the Mutsu Ogawara port, only kilometres from the nuclear facility. Local opposition to the escalating nuclear programme has dubbed Japan a 'flat without a toilet', because despite having over fifty nuclear power plants, there is still little idea of how to deal with the waste. Two days later, after an emergency cabinet meeting between the Aomori governor and the country's Science and Technology Agency, the *Pacific Pintail* was finally allowed to dock. After many hours, the governor managed to extract a letter from the agency stating that it could not and would not make the province of Aomori the final destination for the waste without his consent. As a shrewd exercise in political arm-twisting, however, Morio Kimura's action has a precedent in the state of Idaho.

The phone rang. 'Cees Andrus?'

'Oh, hi Gov.' After a short conversation, the governor of Idaho put the phone down. He needed a little advice from Cecil Andrus, his predecessor. Although political opponents, they are good friends – and the new governor had not yet found his feet. 'I sneak in the back door, now and again, and give him advice,' Andrus explained. As you might expect from a former logger, Andrus's political advice comes from the fast-talking, 'no bullshit' school of politics. It's a style that is gaining popularity in the rest of the country. The new Republican governor could, apparently, use a little. Andrus's Wild West approach has been particularly effective in Washington DC where he openly displays his contempt for politicians who hide what's left of their integrity behind 'fancy talk and a chauffeur-driven car'. His retirement as governor in 1994, at the end of his fourth term in the job, was viewed as a big loss by the people of Idaho. Everyone I met in Boise, the state capital, knew 'Cees'; even his political sparring partner admitted graciously that he was a shrewd politician. Andrus knows he's popular. He's a big-head with bravado, which is precisely how they like it in Idaho.

Andrus's office, on the seventh floor of the tallest building at the Boise State University campus, is small, neat and characterless. His only contribution to the décor are a number of British hunting scenes on the walls. He goes deer-stalking in Scotland every year. 'If it's got fur, feathers or fins, I'll chase it,' he says. 'And nuclear waste?' I enquired. Which category did that fall into? 'Oh, fins,' he laughed. 'Those spent fuel rods have cooling fins.'

If you follow Snake river from Boise east to Idaho Falls, you meet the largest of the Department of Energy sites, the Idaho National Engineering Laboratory (INEL), which occupies 2,300 square kilometres of the Snake river plain. Both high-level and transuranic wastes are stored here. INEL is currently the largest storage facility for transuranic waste: in the United States 122,000 cubic metres is kept above and below ground. Andrus became aware of the problems of nuclear waste when he was first elected governor in 1970. On taking office he discovered that the state had been given an assurance in the 1950s from the Atomic Energy Commission (AEC) that INEL was only a 'temporary' storage facility. The same year, having undertaken numerous characterization studies, the AEC recommended a salt mine in Lyons, Kansas, as the final repository for high-level waste. The following year its plans were abandoned after a team of independent surveyors published a report claiming that the mine had serious geological flaws that had been overlooked by the government experts. On hearing the news, Andrus wrote to Dixie Lee Ray, the Secretary of Energy, stating his dissatisfaction with the amount of waste stacking up in his state. He vividly recalls her reply. 'She wrote back in a very patronizing manner. "Dear Governor, don't get excited," and she sort of patted me on the bald head. She implied that she was much smarter and knew much more than me. Well, I don't take very well to that kind of treatment.' The letter marked the beginning of a long and dramatic battle over the storage and transportation of waste in the state of Idaho. In her letter Dixie Lee Ray gave Andrus assurances that all the nuclear material would be removed by the end of the 1970s. The waste remains in Idaho to this day, however, and continues to increase.

'The world has ignored nuclear waste,' Andrus told me. 'It puts money and sex appeal into bigger bombs, faster reactors, and it allows all the waste to pile up on back lots across America. Right now, Idaho has 9 million litres of sodium-contaminated high-level liquid waste from Rocky Flats and the Navy, housed in thirty-year-old tanks. That's just a time bomb waiting to go off.'

In 1976, Cecil Andrus left Idaho for Washington DC to join the Carter Administration as Interior Secretary. He returned when Ronald Reagan was elected and later went into private business. He admits that he didn't pay much attention to the nuclear-waste issue until he was re-elected as governor in 1986. Back in office, he soon discovered that nothing had changed: it was business as usual. When the Government's plan for permanent storage, the Waste Isolation Pilot Project (WIPP) in New Mexico, failed to open on the proposed date in 1988, Andrus took a drastic measure. 'I figured somebody had to do something, so I closed the state borders and did a little hooplah,' he recalls. He also notified the Department of Energy that any shipments of nuclear waste into the state would be returned to sender. In order to prove that he wasn't playing games, Andrus deployed the police on the state line and ordered the National Guard to be on full alert. 'All the legal advice I had said, don't do that Governor, the state doesn't have the authority. Well, I told them that they worked for me so they should go back to their offices, keep their mouths shut and I'd handle it. I told the National Guard to get an M60 tank ready, and I personally called the Union Pacific Railroad who ship the stuff in and told them that they had a serious problem on their hands. Of course, they ran for cover immediately. I used some bombastic type of rhetoric like, "I'll sling the muzzle of that turd right down that track and we'll see who blinks." Out of character for me, but I had to get the people's attention.' Attention was exactly what Cecil Andrus got. News that a train carrying waste from the Department of Energy's Rocky Flats plant had been stopped in Blackfoot, Idaho, just before it entered INEL was reported in *The New York Times* the following morning. Accompanying the article was a photograph of an Idaho State Trooper on the railway track, with his arms crossed in front of his chest. 'It was the perfect picture,' Andrus remembers. 'This trooper had biceps the size of my thighs. The article talked about some hard-headed and bald-headed governor out in Idaho. It was the first time the issue got national coverage.' And then, mellowing, he says, 'Somebody had to do something. The world has got to realize that this is a very serious issue. I had been elected to represent the people of Idaho and was therefore responsible for their safety and protection. That waste is stored on top of the largest freshwater aquifer in the US, the Snake River Aquifer. Ninety per cent of south Idaho's population gets its drinking water from that aquifer, in one way or another. Also, we're known as an agricultural state; can you imagine what would happen if radioactivity showed up in the water being used to water food crops? We wouldn't be able to sell a single, god-damn potato. People would be scared to death.'

It was three o'clock in the morning. Daniel McKay had been driving for eight hours and had stopped just west of Buffalo, near the state line between New York and Pennsylvania. He couldn't understand why he hadn't yet overtaken the train or even caught a glimpse of it. Was there another route? He had followed the most logical route on the Conrail line, from Albany through Schenectady, Syracuse, Rochester and Buffalo, and then south to the Norfolk and Weston line. He parked the car just outside Fredonia, within sight of the junction of the two railroads. He felt that he was bound to see the train here. All he had to do was stay awake. In the back of the car were enough supplies to keep him going for the whole journey, a seven-day chase through ten states of America to the Idaho National

Engineering Laboratory. He had planned the enterprise for nearly two years, and yet he had lost the train on the first day. He had considered all the possible sidings it might sit in for a few hours just to throw him off the trail. He knew all the possible connections and routes. He had a list of all the radio frequencies the rail workers used to communicate with the caboose. He must have been close. Why else would he have been stopped by the state police wanting to see his ID, vehicle registration and licence? They had taken a long time checking his credentials, even though he hadn't broken the law. The policemen later admitted that they were obeying orders from the FBI.

The overheated engine kept the car warm for an hour or so. October nights in New York State can be pretty cool, and by 4.30 a.m. it was downright cold. Nevertheless, McKay fell asleep. By morning, he was tired and stiff. He recalled waking each time a train rattled past, but it was never the *right* train. Had he missed one? What could have happened? He resolved to continue the journey to a meeting place on the state line where he had arranged to pick up Anne Sorenson from Nukewatch. They had spoken a number of times on the telephone and she had impressed him as this very bright, savvy young woman, so he was surprised to discover that she was in fact in her sixties. Together, they decided to retrace his steps. Back in Albany the following day, they checked with the groups positioned ahead of them, to see if anybody had seen the train. Nobody had. They waited all day outside Albany parked next to the only railroad track. But they saw nothing. After waiting for three days they decided to change tactics. They started to check the sidings in and around Albany and Schenectady. In Saratoga Springs, in the middle of nowhere, McKay and Sorenson at last found the two elusive casks. On the fifth day, a third cask was loaded and coupled to those already waiting in the siding at Saratoga Springs. By late afternoon, the cargo started to move. Sorenson and McKay gave chase. They stopped momentarily on a bridge to see the train pass underneath them, down a single-track corridor leading through the forest. The nukewatchers drove parallel to it as it entered Schenectady and disappeared into the town centre. The local police were also in pursuit, but McKay ignored the wailing sirens and flashing lights. North of Schenectady he parked and waited for the train to emerge from the town on its way to Syracuse. But it never showed up. In fact, as McKay and Sorenson later discovered, the nuclear train had switched on to a track heading south towards Binghampton.

The US Navy transports its spent nuclear fuel by rail across the country from eight naval ports and research centres on the east and west coasts. Since 1957 there have been some 600 shipments of highly radioactive waste. The Navy protests that its shipments only contribute to 2 per cent of the military's high-level waste but, as McKay points out, over the next forty years its share will rise to nearer 50 per cent. McKay runs the Knolls Action Project (KAP) from a small first-floor office in downtown Albany that he shares with Campus Action, the Coalition Against Apartheid and Act Up. He regards it as a 'humble little project', but in fact his work has led to the halting of the Navy's nuclear-waste shipments. McKay's office, a computer and a phone surrounded by filing cabinets, is the centre of the KAP operation. Tables, charts, rail maps and timetables are neatly stacked in the corner of the room. What angers the Navy is being unable to stop its clandestine operation becoming a media event: apart from the occasional paint bomb, McKay's actions are perfectly legal and – some would argue – necessary. Until he began to photograph the shipments, noting the times of arrival, departure and regular routes, local authorities had no knowledge of radioactive waste travelling through their territory and were consequently unprepared to react in cases of emergency. When a truck carrying naval torpedoes overturned at five in the morning on one of Denver's busiest highway intersections in August 1984, the emergency services had received no prior warning of a nuclear truck being in the area – it wasn't even scheduled to be in Colorado.

The KAP was set up in 1978 to raise awareness in the community about a naval research station, Knolls Atomic Power Laboratory, that is hidden in 1,600 hectares of pine forest near Saratoga Springs, north of Albany. McKay took over the running of the project in the late 1980s, just as it was about to fold. Instead of chaining himself to wire fences or spray-painting the roads in protest, he focused his efforts on what he regards as the Navy's Achilles' heel, the transportation of spent fuel. He decided to establish a network right across the country that could publicize and follow the secret shipments from the laboratory to Idaho. 'Little or nothing was known about these shipments,' he recalls, 'despite the fact that they use a rail network designed to go through urban areas, such as Cleveland and Kansas City. Until then, nobody had photographed them, inspected them or knew when they travelled. The Navy has never given out information about the shipments willingly. It has been outside public control for so long that it still has a Cold War mentality.'

According to McKay, you always know in advance when a shipment is about to leave the Knolls Laboratory. Two Portakabin toilets are placed in the dirt track outside the elementary school at Ballston Spa. The track is in fact an old railway siding that leads to a derelict rail yard. It is here that the Navy loads its spent fuel casks on to low loaders, which are then coupled to an engine and pulled to the siding in Saratoga Springs. A usual load consists of two or three casks. Each takes a day to load – hence the toilets. Another useful index is the removal of the low overhead cables in Ballston Spa's main street. The large trucks that transport the casks from the laboratory to the siding go straight through the middle of town. 'It's a pretty exotic event,' says McKay. 'Everyone comes out to stand around and watch. It's like a 4th of July parade.' He and his fellow protesters make sure they get ahead of the convoy in order to herald its arrival with anti-nuclear slogans and banners. 'We get ignored most of the time. A lot of people in this community are very complacent. It's a sad thing to say, but it would take an accident to shake them up. The one thing in this country that everyone has an opinion about is sport. In sport there is a clear divide: you have a winner and a loser, and that's how you know whether one team is better than another. In this debate, however, the issues are grey and difficult. People end up not knowing what to think. They get *confused*.'

Following a train is a lot more difficult than you might imagine, especially if it doesn't want to be followed. The rail network outside Albany is a complicated web of tracks and sidings, half obscured by pine forest. There are numerous routes a train can take, and each time it goes into a station you can lose it simply by panicking. The key, according to McKay, is a combination of patience and a belief in your own judgement. There are, of course, a number of tricks. The radio scanner is a useful tool. Hardcore train-spotter magazines print the radio frequencies used by the different railroad companies, so that enthusiasts can listen to conversations between the train drivers and signalmen. Another device McKay has made use of is the computerized information service, a service that allows you to dial a 1-800 number, punch in the reference number of a caboose and find out where it is. 'If you can find out where the caboose is you know exactly where the waste is,' says McKay. 'The Navy only has six cabooses and all they do is escort nuclear casks. Each caboose has two armed guards on board and a sophisticated satellite tracking device.' Unfortunately, a year ago the rail service got wise to McKay's ploy and discontinued the service for naval trains. So he tried the same thing using the reference numbers of the goods trains that travel with the casks. But again, the service eventually caught on to what he was doing. The next thing he did was to cultivate 'moles' in the train service, sympathizers who call now and again with anonymous messages about where and when a shipment is leaving. Others send him classified information regarding safety standards or the testing of the casks.

Cecil Andrus feared that his actions in 1988 would get him into a lot of political trouble. Others argue that he knew damn well what he was doing: in 1990 he was re-elected for a third term as governor of Idaho with a record 91 per cent of the vote. 'After that, hell, no one could beat me,' he boasts. 'I had no legal right to do it but the Feds blinked, and the minute they blinked I had 'em. Stupid as they are, they entered into a memorandum of understanding with me, and as soon as they did that I had some legal standing. They gave me the club to beat them with.'

In January 1991, Andrus wielded the club again. He banned road shipments of spent fuel from a nuclear power plant that was being decommissioned in Fort St Vrain, Colorado. In defiant mood he notified the press, saying, 'I've got a state policeman and fifteen of his friends at Blackfoot, and all of them are prepared to do what is necessary if that truck makes it inside the borders of this state.' Admiral Watkins, the Secretary of Energy, and Leo Duffy, his right-hand man, tried to pacify Andrus by saying that the waste would be stored temporarily and was only being used for 'research' purposes. Andrus responded by asking them to sign a letter guaranteeing that if the waste was not used for research, it would immediately be shipped out of the state. 'The Department of Energy has nice guys in leadership roles, people like Mr Watkins and Mr Duffy,' the governor told reporters. 'But we've dealt with nice guys before,' he said at the time. 'What I've asked them to do, besides being nice guys, is to put in writing what they've agreed to do verbally.' The Department of Energy refused his request and the shipments from St Vrain went elsewhere. Suits and countersuits went back and forth between the utility running the St Vrain plant, the state of Idaho and the DoE. Towards the end of the year, Idaho was granted an injunction to keep further shipments of civil spent fuel out of its territory. The injunction, however, was lifted six months later. In October 1992, Cecil Andrus requested another injunction, blocking shipments of the Navy's nuclear waste that were coming into the state at a rate of one cask a week. On 28 June 1993 the federal court agreed to his request.

In December the Navy appealed to Governor Andrus to allow a limited number of shipments into Idaho, on the basis that national security was being jeopardized. The ships and submarines, it argued, could not be refuelled, because there was nowhere to store the spent fuel. Andrus agreed – and over the next year nineteen shipments made it to INEL. Just before he stepped down from office, Andrus refused to allow an additional eight loads, so the Navy reapplied to his successor, a Republican called Phil Batt. The new governor immediately acceded to the request, even though 80 per cent of the population of Idaho, it was reported, was opposed to the decision. 'If I had to do it again, seeing the political capital that you lose by attempting to do the thing you think is correct, I probably would not have done it,' a remorseful Governor Batt told the *Idaho Statesman*. 'I believe the mood of the people in Idaho is not to accept any shipments. I represent them. I share that concern and I will resist more shipments, vigorously.' On 17 October 1995, the state of Idaho, the Navy and the Department of Energy agreed to allow 1,133 shipments of spent nuclear fuel into INEL over the next forty years. The decision was approved by the court. According to Beatrice Brailsford of the environmental group Snake River Alliance, the agreement is an appalling insult to the eight-year struggle for the protection of Idaho's population and environment. She says that most people knew that the Batt Administration was going to fold under the pressure. 'Congress amended one of the appropriation bills to force us to take naval waste for national security reasons,' she explains. 'But the number of shipments doubled, from twelve to twenty-five, without the knowledge of the Defense Department.'

Not long after the agreement, a Pacific Union train was carrying nuclear waste across the Fort Hall Indian Reservation to INEL. The driver saw a line of red flares burning on either side of the track ahead. He eased on the brakes and signalled with his horn that a train was approaching. In the distance he could see that a tribal police car was parked across the tracks and that the area was surrounded by Shoshone-Bannock police and tribal leaders. The train ground to a halt and the police mounted the engine. The tribe had been notified of the arrival of the shipment by Daniel McKay's network. The train was held up for six hours and then escorted off the reservation by tribal police. The Shoshone-Bannock are now in negotiations with the Navy and the Department of Energy.

'The Government has found it easier to say "that's top secret" than to explain anything to the public,' says a defiant Cecil Andrus. 'Hell, there's nothing top secret about a nuclear-powered reactor, but it is easier to keep the great unwashed out there, uninformed, then they can do as they please. I don't think you will see a permanent repository for high-level waste in America in your lifetime or mine. I hope I'm wrong, but I don't think so. I'd be willing to let them bring in a pound if they take two pounds down to WIPP (Waste Isolation Pilot Project) in New Mexico, where it belongs. I think I'll see that happen. I might not be out there in the forefront fighting, but I'll still be a player.'

Fred Ferebee lives in an old signalman's cottage on Platform One at Penge East railway station. For a committed train lover like Fred,

Station Cottage is ideal. When he moved in, he spent a lot of his time watching the trains from the vantage point of his spare bedroom window. (The room is now used at night by the police on the lookout for vandals.) 'We've got no toilets on the platform because they kept smashing them up,' bemoans Fred. When I arrived at his cottage, he had prepared an immaculately presented lunch of bread, pickle and cheese, garnished with grapes and paper napkins. It would be inaccurate to brand Fred a 'train-spotter'. There wasn't an anorak or acrylic pullover in sight. Fred, in matching green shirt and trousers, was as neatly turned out as his lunch. His cottage is small and tidy. He has eccentricities, though. Instead of activating his new £5,000 alarm system whenever he goes out, he switches on the radio and places his spare glasses and cigarettes on a table within easy view of the window. When Fred bought the cottage fourteen years ago, the estate agent had neglected to tell him that the nuclear trains pass only a metre from his front door en route to Sellafield from Dover and Dungeness. 'If I'd known that, I think I would have had second thoughts about buying the place,' he told me. When watching from his spare-bedroom window, he noticed occasionally a goods train carrying a nuclear flask pass through the station at rush hour, as commuters lined the platforms. Alarmed, Fred brought the matter up at a local CND meeting, and within a week he was contacted by the London Nuclear Trains Working Group, who asked him to keep a note of the appearance and frequency of the trains. Fred is now the group's Penge activist.

Trains carrying highly radioactive waste through Central London have become legendary. They are usually referred to as 'nuclear' trains, or 'white' trains. The first time I saw a nuclear train I was in the company of Fred Ferebee. It was a Thursday afternoon. After a long tramp through wet undergrowth, the two of us eventually found the train abandoned in a siding at Hither Green Station. It sits there every Thursday for three hours waiting for a change of crew. On sighting the train, we began to speak in whispers as if we were stalking it, like big-game hunters – in Sydenham. Yet the train was disappointingly nondescript. It looked docile and distinctly unthreatening. A small radioactive sticker on the steel container was the only indication that we were looking at the right thing. Fred scribbled the chassis number in his notebook and deliberated over whether it was an AGR or Magnox flask. We took several photographs of the train and inspected it thoroughly through the binoculars. Paul Hawkes, a CND spokesman, discovered that the train was due for a brake test in August and that it had come from Depot 43, probably Dungeness, which led Fred to believe that it was in fact an AGR flask. The life of a 'nuclear' train-spotter involves painstaking hours of waiting for a train to appear, and when it finally does, nothing out of the ordinary happens. After several outings you become interested in the smallest changes and the minutest details. You begin to believe in conspiracy theories. Are you looking at a dummy flask? Has the real one been re-routed in order to avoid detection? Do the letters and numbers on the side of the container provide a vital clue to illicit activities? The Nuclear Trains Working Group has compiled a dense list of timetables, routes, appearances, numbers of flasks and flask numbers, which it uses as a sort of reference.

Trains pass through London every Thursday carrying 2 tonnes of irradiated nuclear fuel in each steel flask. Each flask holds approximately 200 to 300 spent fuel rods (400 fuel pins in an AGR flask) or 80×10^{15} becquerels of radiation, 'Whatever that means,' Paul Hawkes observed. The trains come from the nuclear reactors at Sizewell, Bradwell and Dungeness on the Suffolk, Essex and Kent coasts. All the flasks meet in a siding at Willesden Junction on Thursday evenings, where they are coupled together and continue their journey north to the THORP reprocessing plant at Sellafield. Until recently spent fuel from Germany, Italy and Switzerland was shipped to Dover, where it continued its journey north by train, passing Fred Ferebee's cottage. Fred would receive a call from somebody in Dover who had been tipped off by a man in Dunkirk about a shipment that was on its way. As it went past Fred's back door, he noted the time, number of flasks, approximate speed – the trains are required by law to travel no faster than 65 kmph – and the flask number, if possible. He then called the next activist up the line, and on and on it would go through Brixton, Clapham Junction, Willesden Junction, all the way to Cumbria.

Early in the morning on Friday 7 July 1995, a train stopped at Rugby station for a change of crew. To the passengers standing on the platform it was just a goods train pulling five low-loaders with a guard's van at the rear. Three central carriers supported large, ribbed, steel containers. At half past seven, however, the train, already three hours late, was heard 'crackling' whilst standing opposite Platform One. A commuter who heard the noise notified railway staff and the train was pulled into a siding. The crackling continued, and at 8.32 a.m. the police were called. Six minutes later the alarm was raised and the local fire brigade arrived in protective clothing, with full face masks, breathing equipment and Geiger counters. The station was closed and a 50-metre exclusion zone was set up around the train. Police in black overalls were brought in to reinforce the no-go area and to clear the footbridge over the track, because people had gathered to watch the activity. The passengers already in the station were not evacuated. One of them overheard a police constable saying that one of the flasks was leaking. The locals, whose houses back on to the siding, were told to stay indoors and to keep their windows closed. At 10.37 a.m. a team of health physicists arrived from the nearby Rolls Royce plant which makes reactor plates for nuclear submarines. By 12.45 p.m. a second team of health physicists arrived from the Oldbury nuclear reactor in Gloucestershire.

The official response from the Department of Transport highlighted a problem with one of the engines, hence the lateness of the train. It also added that although flask E83 was being investigated, the radiation levels were normal. ('Normal' means within the acceptable but nevertheless detectable levels.) It was unclear exactly what the problem had been, however, and the train later continued on its journey to the British Nuclear Fuels' reprocessing plant on the Cumbrian coast.

According to the Nuclear Trains Working Group, radiation from a flask carrying spent fuel rods can be detected from anywhere up to 600 metres away, depending on the type of reactor fuel. The new housing estate adjacent to the Hither Green siding was inside that range. Residents who live within 600 metres of Rugby, Stratford, Willesden Green and a host of other stations where the flasks are stationary number in their tens of thousands. Health physicists argue that there is no health effect from living that close to the line or from being exposed to such low levels of radiation. But they only mean that such exposure has no significant *detectable* effect upon the health statistics of the overall population.

At sixty-six, Fred Ferebee is now a senior member of the unofficial nuclear train-spotters' network. As such, he helps his colleagues to monitor the trains, distribute leaflets and organize protests. In 1994 Fred was branded a 'scaremonger' by BNFL after he masterminded demos in Penge, Bromley and Brixton. 'About forty of us, dressed in white overalls and gas masks, leafleted shoppers in the high street,' he recalls. 'The reaction was generally very supportive. I think the only person who told us to stop interfering was a local Conservative councillor.' Nevertheless, Fred feels uncomfortable with the label 'activist' or 'nuclear campaigner'. Two CND banners from the Bromley demo propped against the wall of his kitchen are mementoes rather than the tools of his trade. The paint pots in his study are not intended for writing political graffiti, but for his other hobby, water-colouring. His sitting-room, a train enthusiast's haven, is decorated with railway workers' lanterns, goods vehicle registration plates and ticket machines. The books, hardly subversive tracts, recount the history of the steam age. 'As a kid I used to go to Victoria Station and watch the big steam locomotives,' he told me. 'To me they were friendly things. There was nothing threatening about them. It's ironic really.' Not least because Fred's knowledge of trains, of their markings and movements has proved extremely useful to the London Nuclear Trains Working Group, most of whose members are nuclear campaigners first and train-spotters second. At the same time Fred has benefited from his involvement. He took early retirement several years ago. Now, divorced, with grown-up children, he finds that the campaign has added a new dimension to his life. The last time we met was at Snaresbrook Crown Court. Along with about forty other campaigners, mainly CND members, Fred was taking part in a demonstration to demand the acquittal of Pat Arrowsmith and David Polden, who were being charged with holding up a nuclear train in April 1995. The demo was also intended to raise awareness of the nuclear trains in Snaresbrook. It was a sedate affair. Leaflets were distributed and banners were draped over the railings at the entrance to the court. The average age of the protesters was over fifty. There was no shouting or chanting, although people did clap, mainly for the local news camera, when Arrowsmith, wearing a waistcoat emblazoned with CND logos, sauntered down the road. 'Free David Polden,' somebody called out, and then we filed into court for the hearing.

Pat Arrowsmith and David Polden were arrested for holding up a nuclear train for twenty minutes by sitting on the line at Platform Ten, Stratford Station, in April 1995. Arrowsmith, now in her mid-sixties, is a veteran peace campaigner and CND's main spokesperson. According to reports, she leapt on to the track in front of the Bradwell train, as it waited for a crew change. 'Oh God,' the driver was heard to exclaim, 'not now, I want to get home for my tea.' Much of the attention focused on the sit-in, but some demonstrators painted 'DANGER' and 'HELP' on the side of the train. Others, including Fred Ferebee, paraded up and down the platform with banners. The police arrested Polden and Arrowsmith and charged them under the Lethal Damages Act of 1861, which has a maximum sentence of two years' imprisonment. Arrowsmith and Polden were acquitted on a technical matter. It was argued that the case against them had been wrongly constructed, because the prosecution had not stated what the defendants had done to obstruct the train. It was a bad day for the Crown Prosecution whose chief witness, the train driver, failed to show up. Apparently he was tending his allotment in Ockendon at the time.

In 1947 a report by the UK Atomic Energy Authority on the level of radioactivity at Harwell led to a public debate about safe and efficient ways of transporting nuclear waste to a disposal site. The report stated that 'rail transport should be ruled out, because of railway regulations and passing through populated areas.' In 1951, the Berne Union met to discuss the question of transportation. It was proposed that the members of the union could use rail transport but should adopt US regulations. Britain, however, not being a member, decided in isolation to operate its own code of practice. In 1954 a letter to the Canadian Atomic Energy Organization from the research establishment at Harwell stated that, 'For political reasons we have no regulations in this country governing the transport of radioactive isotopes on the scale which is your concern . . .' It is unclear what those 'political reasons' were. A Central Electricity Generating Board leaflet, published the same year as the Chernobyl accident, asked the question, 'What truth is there in the claim that if a flask leaked, large, heavily populated areas, London, for example, could be uninhabitable for decades?' The answer? 'This is an absurd suggestion which has no valid scientific basis.' British Nuclear Fuels Ltd claims that it hasn't witnessed a radiation leak in thirty years of operating the trains. The flasks are tested according to international standards derived from the nuclear industry's assessment of a 'credible' accident, as opposed to an 'incredible' one. In the 'drop' test, a 45-tonne flask is dropped from a height of 9 metres on to an 'unyielding target'. According to the nuclear industry, 9 metres is the height of the average bridge, but Fred Ferebee says that this fails to take into account the 30-metre drop from the viaduct at St Mary's Cray near Bromley. The second test is the fire test. A flask must remain intact after being engulfed in a fire at 800°C (the temperature used for testing safes) for half an hour. According to John Large, a nuclear expert, the bolts on the flask are tightened to such a degree in order to pass the first test that they would cause it to fail the second, because the flask would not be able to release the internal pressure by expanding in the heat. The third test is the immersion test. A flask is immersed in water for a period of eight hours. But again, as Large points out, a recovery operation from the Thames (the Dungeness train travels across the Thames) would take longer than eight hours.

These criticisms might appear pedantic. At the beginning of 1995, however, it was announced that the number of staff monitoring the transportation of radioactive materials was to be cut. Since then a 50-tonne flask has been dropped twice in the course of being loaded from a lorry on to a train at Sizewell. Witnesses claim they saw the flask stuck in the ground, corner first, and the gantry crane at an unusual 40° angle. In February 1995 another train carrying two flasks from Sizewell was stationary for over an hour near Ingatestone because of an overheated axle. This fault can lead to a derailment. The previous year, the derailment outside Maidstone of a train carrying cargo from Germany led to speculation that it was also carrying a nuclear flask. Not long afterwards there was a derailment at Hartlepool nuclear power plant due to points failure. Teenagers placed a concrete sleeper across the line in the Penge East tunnel, almost derailing a passenger train from Margate. A similar object laid across the tracks near Clapton, en route to Sizewell, caused a train to derail in the mid-1980s. There have also been a number of fires recently on the North London line, the main route from Sizewell and Bradwell. One of the fires held a nuclear train up for five minutes at Gospel Oak Station where there were people standing on the platform only metres away from the flask. A passenger was told by a British Rail worker that there was no need for concern as it was only radioactive carbon! At South Woodham Ferrers, on the Bradwell route, there is an unmanned level crossing. Until recently the crossing was part of a busy road link into the town centre. According to Peggy Owen, who lives next to the crossing, a traffic accident would result in a derailment and a train ploughing through houses that line the track. The crossing was unmanned because BR planners, working from an old map, were apparently unaware of Woodham New Town's existence, in spite of there being a station there.

Fred Ferebee and other unsung heroes of the railway sidings battle against public apathy, media indifference and the powers of bureaucracy. Even so, after thirty years of near-misses, the future of nuclear transport by rail is beyond question. What is so admirable about Fred's work is the way he carries on regardless, on cold winter evenings, waiting for the trains to appear. He never tires of handing out leaflets, dressed in white overalls and gas mask, or of marching up and down station platforms holding a CND placard. It would take that 'absurd suggestion which has no valid scientific basis', a major release of radioactivity into a densely populated area, to end the practice once and for all. But in such an event, I doubt that Fred Ferebee would take any comfort in saying, 'I told you so.'

10. *Atomic Dustbins*

A hundred kilometres north-west of Las Vegas, in a remote and windswept desert, the Yucca mountain rises 1,800 metres above sea level. Temperatures here reach 40°C during the day and drop below freezing at night. Only 80 km to the west of Yucca mountain is Death Valley, the lowest point in the hemisphere. As the sun recedes behind the mountains in the west, the desert begins to resemble the surface of the moon, scoured and scratched. This silent landscape can be read like a big geological map. From the top of Yucca mountain, the eye can spot evidence of the Miocene and Palaeocene eras, volcanic structures that have ceased to erupt long ago, ancient rivers and ice flows that are no more than marks on the desert floor, as if a piece of wood had been dragged in the sand. At the base of Yucca mountain, beneath a ridge called Exile Hill, the Department of Energy (DoE) plan to bury up to 70,000 tonnes of radioactive waste from America's 110 nuclear reactors and twenty-four military sites. The permanent repository will be constructed 300 metres below the surface, in thick volcanic rock, 250 metres above the water table. When the Nuclear Waste Policy Act was passed in 1982, there were nine potential sites for deep geological disposal. Three were finally short-listed in 1986: a salt dome in Deaf County in Texas, the basalt rock at Hanford in Washington State and the volcanic tuff at Yucca mountain. In 1987 Senator J. Bennet Johnston, a Democrat from Louisiana, amended the Nuclear Waste Policy Act to restrict competition for the repository to Yucca mountain and offered Nevada a benefits package of US$20 million a year if it waived its rights to disapprove of the siting. In some circles the amendment became known as the 'Screw Nevada Bill'; it was felt that the choice had been made because it offered the path of least resistance, both politically and publicly. In 1987 the geological exploration of Yucca mountain began.

The mining law hasn't changed since it was first drawn up in 1872. The law states that a member of the public can prospect on any public land and, if they can prove that the land contains minerals, they are within their legal rights to stake a mining claim to that land. One claim represents a plot 450 metres by 180 metres. The prospector makes the claim by driving stakes into the four corners of the plot and placing a small square white flag in the centre of the land. The claim must be filed with the state, the county and the Bureau of Land Management. Maps have to be drawn up and the plot must be surveyed. In most parts of America the law receives scant attention, but in southern Nevada prospectors still make a good living by staking claims and leasing them to mining companies for anything up to US $20,000 a year. A skilled prospector will negotiate a deal whereby the mining company will buy him out, sometimes at a cost of millions of dollars, if it finds what it's looking for.

Two hundred kilometres north of Yucca mountain, Tonopah is a small but important town in this remote part of Nevada. It is built on the intersection between two main highways: Route 6, heading east to west, and Interstate 95, heading north to south. Over the years it has attracted a number of mining companies on account of the gold, silver and copper in the surrounding land. One way or another, most of the people in Tonopah are employed in the mining industry. The gold-rush fever lives on, and most of the conversations in the bars end up returning to the time-honoured dream of becoming rich by prospecting for gold.

Anthony Perchetti runs a construction company in Tonopah. As a subcontractor for the mining companies, he excavates land and caps drill sites. The Perchettis are a well-known mining family in the area. Between 1987 and 1988, Perchetti filed twenty-seven mining claims on and around Yucca mountain. 'I knew that there were minerals in there,' he told me. 'There's a gold mine not 30 km from Yucca mountain. At the time, Bullfrog Mining were drilling about 9 km from the mountain, and I knew they were interested in making their own claims on the land, so I just got in there first. I figured that if the mining company didn't buy the claims, then the Government would. One way or another, I knew I had a buyer.' Perchetti staked his claims after gaining access to Yucca mountain through the nuclear test site. As he was up there driving pegs into his plots, he was approached by government geologists who were also surveying the land. They asked indignantly what he thought he was doing. 'I told them I had every right to be up there as I had filed claims on the land,' Perchetti recalls. 'It was all legal. They went away with their tails between their legs to check it out. You see, the Department of Energy made a mistake. They should have withdrawn that land from public use before they even started talking about using it as a waste site. All they had to do was go to the Bureau of Land Management. It's a real simple thing to do.' A few days later Anthony Perchetti was contacted by a government lawyer. He invited him to a meeting at the Department of Energy's head office in Las Vegas to sort the problem out. At the meeting Perchetti was introduced to Carl Gertz, the manager of the DoE's Yucca Mountain Project and its chief negotiator on the case. 'He seemed like a nice guy, very professional. He brought in four or five lawyers and they did everything they could to make me give it up. But I stood my ground. I knew I was in a good position. I had filed on the land and they would have to make me an offer. I was hoping to get around US$2 million.'

'I think we had a good case,' Carl Gertz told me confidently, 'I knew that if we wanted to take it to the courts we could win our case. There hadn't been any mining in that area. Mr Perchetti was aware of the loopholes in the law and he knew that the land hadn't been withdrawn formally, so he went ahead and exploited the situation. There's a strong entrepreneurial streak in this state, particularly in the mining community. People often make what we call "nuisance" claims. It's simply taking advantage of an old mining law that hasn't been amended.' Perchetti denies the allegation that his claim was a 'nuisance'. 'A nuisance claim is where you might stake a claim on the land surrounding a working mine in the hope that the ore goes beyond their boundary, and they then have to buy you out. My claim wasn't a nuisance claim, because I had papers from geologists stating that there were minerals there.'

The Nevadan entrepreneurial streak was displayed soon after Yucca mountain was nominated as the waste site. The Nevada State legislature took the unprecedented step of inventing a new county. Three hundred and seventy-five square kilometres of desert around the Yucca mountain was named Bullfrog County in order to take advantage of the Grants Equal to Taxes (GETT) clause in the Nuclear Waste Policy Act. The clause stated that the repository was liable to a form of real-estate tax. Not surprisingly, Bullfrog County had Nevada's highest taxes, and because nobody lived there, the money was automatically paid into the state's coffer.

The claims staked by Perchetti posed a problem for the Department of Energy because the disputed land lay right over the top of the proposed repository. 'In order to do our job properly we had to eliminate all interference with the site and make sure that the physical integrity of the subsurface was not altered in any way,' says Gertz. In Las Vegas, the negotiations about payment continued – Perchetti recalls meeting Gertz and his team of lawyers at least ten times – until they reached a stalemate. Gertz made a final offer of US$249,500. 'I tried to explain our position to him and I basically told him that he could either take it to litigation or we could get him a cheque in a couple of days.' Perchetti knew that taking the Department of Energy to court was expensive and would take several years to settle, so he accepted the offer. 'It was a pretty unique situation and we were happy with the settlement,' says Carl Gertz. 'Two hundred and forty-nine thousand was an acceptable figure. Perchetti could have made life awkward for us. We could have stopped him getting mining equipment on to the land, but he could have got a drilling rig up there using a helicopter.'

Anthony Perchetti has ten other claims in the area that he is trying to lease. 'I've had some good claims for over twenty years on old gold mines close to Yucca mountain. But it's more difficult to make a living these days, because the cost of filing a claim is increasing. You have to pay the Government US$100 a year per claim and the county gets twenty dollars a year, so you can't afford to keep too many. I think it's great when a guy like me can go out and stake a claim on a plot of land. It's like the spirit of the Wild West, and hopefully it will stay that way. It would be a sad day if they changed that law.'

> The land was not willed to you by your ancestors
> – it was loaned to you by your children
>
> Kenyan saying

There is a sense of urgency in the nuclear waste industry at present. It is an urgency driven by two main desires, a moral decision not to bequeath this poison to future generations and the hope that finding a solution to the waste problem will give the moribund nuclear industry a new lease of life. The solution now widely accepted by the industry, and by the relevant government agencies, involves burying the waste in underground sites. In the West sites are being explored and built at enormous cost. Deep geological disposal has a neat symmetry to it. The anthropologist Mary Douglas is credited with defining waste as 'matter out of place', in other words, returning to the earth what originally came from it seems profoundly logical. It also complies with a common desire to keep waste out of sight and out of mind. Everything from shooting nuclear waste into space to burying it between the ocean plates in the mid-Atlantic has been considered.

Are we creating more problems than we are solving by opting for permanent storage? Two main questions need to be answered, with confidence. How can we guarantee that the waste will not interfere with mankind during its radioactive lifetime, a period lasting millions of years in the case of some isotopes? How can we make sure that mankind does not interfere with the waste? The first question is already being extensively researched. It has been decided by the US Environmental Protection Agency that 10,000 years is a period of time for which 'meaningful calculations' can be made about the radioactivity of the waste, the geological stability of the dump site and climatic changes. With that figure in mind seismologists, geologists, hydrologists and climatologists, working for the sponsors of the sites, have been studying the past and making cautious predictions. Dr Rolf Meyer, of the Gorleben Salt Dome Project in Germany, argues that the site has been characterized to such an extent over the last fifteen years that the project is being held up by a deluge of statistics, analysis and recommendations. Yucca mountain has had its share of controversy. A furious debate about the groundwater has been going on since 1984. Jerry Szymanski resigned from the Department of Energy's project team because he could not agree with the other geologists that calcium deposits in one of the trenches were due to rainwater seeping down from above. He believes that it is evidence of ancient ejections of pressurized groundwater, and that it could happen again. If it did, he argues, the waste site could be flooded and the groundwater contaminated for ever.

In 1992, an earthquake measuring 5.6 on the Richter scale 20 km from Yucca mountain caused tens of thousands of dollars' worth of damage to the Department of Energy's Operations Center. Anti-nuclear protesters observed that the damage proved the site was unsuitable. Government seismologists, on the other hand, argued that a computer model underlined its suitability by showing that the site had remained stable during the quake.

More recently the controversy has focused on the research of two scientists at the Los Alamos National Laboratories, Charles Bowman and Francesco Venneri, who claimed in *The New York Times* that the waste, 'could erupt in a nuclear explosion, scattering radioactivity to the winds or into the groundwater or both'. The research posed the question of what would happen if plutonium leaked into the rock. Bowman argued that fissile material, having escaped, could reach criticality (a spontaneous nuclear chain reaction) with the rock serving as a moderator to enhance the reaction. As the system got hotter and hotter, and the plutonium dispersed through the rock, more and more neutrons would be moderated and the reaction would escalate. Moreover, if the rock compressed the reaction, it could even achieve supercriticality, at which point its energy doubles in milliseconds resulting in a blast equivalent to thousands of tonnes of high explosive. Colleagues noted that Bowman published his research in *The New York Times* rather than in a scientific journal. It is well known that Bowman favours a solution to the waste problem that employs a technique known as 'transmutation'.

The question of human interference has received scant attention. If we assume that over the next 10,000 years our descendants are likely, albeit inadvertently, to intrude on a nuclear waste site, how do we assess the hows and whys? How do we predict the future? Historians cannot look at the past, as a geologist examines layers of strata, in order to calculate the next 10,000 years. Indeed it is preposterous even to think in such a way. The next generation of nuclear repositories will open during the first half of the twenty-first century. Each site will accept radioactive waste for up to fifty years, after which time there will be a period of custody, or hiatus, thirty years, say, before the repository is sealed permanently, its shafts filled in and the surface works and buildings cleared. The question is how do we alert future generations to the horrors that lie buried underground? How do we prevent a fatal trespass?

A solution proposed by the US Department of Energy involves marking each site with a series of earthworks or monoliths. According to Michael Brill, president of the Buffalo Organization for Social and Technological Innovation, this system 'would pierce the collective unconscious and resonate of death and fear, of a threat to the body. It would attract notice yet viscerally repulse. Like the Holocaust memorial in Berlin, it zigzags, it hurts; it offers neither comfort nor ennoblement. It's not a place where a person feels whole.' In order to explore this communication puzzle, the Department of Energy set up teams with miscellaneous skills, from astronomy to linguistics. The problems were enormous. For a start, the markers had to last the specified 10,000 years and convey decipherable information warning of the location and of its possible threat to a future society.

The first question was: Who are we trying to communicate with? It was agreed that any attempt to predict the future would be misguided. The study considered a range of possible futures. Experts with backgrounds in history, economics, sociology, geography, political science, agriculture and demography were assembled. The ensuing discussion about future societies concentrated on those factors that are thought to influence social activity. Rates of industrial production, density of population, knowledge of the past, economic development, energy resources, the price of minerals and the rate of technological progress – all these factors were taken into consideration. It was thought, for example, that over the next few centuries there might be an increase in the use of solar power to process ores, and that, as a result, there was likely to be a corresponding increase in mineral extraction. It was also felt that government control of any site could not be guaranteed, and that any information held by a government about the site could be lost in the course of political upheaval. A future society might even forget about the site, or it might be aware of the site's existence but forget about the dangers. It was suggested that other media could replace the printed word altogether, and that the existence of a waste site could take on the status of a myth or legend. Cancer might be curable by then, in which case the hazards of exposure to radioactivity would be reduced to some extent. How might a future society intrude upon a waste site? Would technology continue to develop, allowing complex explosions, deep strip or esoteric mining techniques to be deployed in the area of the site? What was the likelihood of robots performing the work? Would futuristic underground tunnels between cities increase the risk of intrusion? Would the repository be accidentally reopened or excavated in a scientific or archaeological dig? 'We toyed with the idea of actually burying some treasure twenty feet down so that anyone digging would think that they had found whatever they were looking for,' says Professor Frank Drake, an astrophysicist from the Lick Observatory in California.

Using the calculations of probability, three scenarios for the future were considered by the team designing the markers. The first described a world in which humanity had regressed to the level of medieval Europe. Technology was metal-based, and there was only limited literacy in an agricultural economy. It was believed that in this scenario the chance of intrusion was negligible. As a matter of fact, the team argued that in the circumstances a marker would become a place of assembly used for religious purposes. The second (and more likely) scenario was a human existence whose levels of technological sophistication continued to vary throughout the world. In this case the likelihood of intrusion was considered to be higher, but it was also felt that a system of markers and messages would be easily decoded. In the third scenario human society underwent a global catastrophe in which it was reduced to illiteracy, a new Stone Age. In this event it was considered important that the nuclear warning be conveyed equally by design and atmosphere. Scientists and engineers addressed the problem of using materials that could withstand vandalism and erosion by weather and pollution. Designers and architects came up with the design and placement of the structures. Archaeologists provided information about past designs and configurations that have successfully remained intact over long periods of time. Anthropologists contributed an understanding about the way humans process information and communicate. Linguists drafted a message, with a view to the way language and meaning have evolved. Experts in semiotics designed signs and symbols to convey the message.

The prototypes for markers varied considerably, but there were also significant similarities. No geometric shapes were used. It was felt that the geometric form gave the wrong impression. (Historically, geometric forms have been used to embody aspirations and ideals.) For the same reasons, no craftsmanship was proposed, because craftsmanship bestows quality and value. The materials used were to be inexpensive and common – concrete and granite, say – to avoid bounty hunters. They should be heavy and awkward to move, 'obnoxious', perhaps containing a bad smell or making an unpleasant sound if caught in the wind. The markers should not provide shelter or any other form of comfort. Different kinds of information were prepared. A series of human faces conveyed torment, fright, anguish, nausea, bitterness and physical pain. In addition, a simple message in all the languages of the United Nations read: 'The danger is in a particular location . . . it increases towards a centre . . . the centre of danger is here . . . of a particular size and shape, and below us. The danger is unleashed only if you substantially disturb this place physically. This place is best shunned and left uninhabited.'

The message was accompanied by symbols – the radioactive symbol, for example, the trefoil with an arrow beneath it pointing downwards, the skull and crossbones. On a more sophisticated level, the necessary information was carved on a double set of granite slabs and enclosed in a concrete room. Access to the room was difficult but possible through a small sliding block of concrete that enabled an individual to crawl in but made it impossible to remove anything. There were to be five such rooms: four buried at the corners of the site, and protected from the elements, one remaining on the surface. Inside the bunker carved messages informed the visitor of the amount of waste, its location, its radioactivity and, using the periodic table, the types of isotope stored. It stated the time of burial in five different calendars, as well as providing a star map. Smaller markers were scattered elsewhere on the site in order to warn of the danger from a distance.

But how 'meaningful' is it to think in terms of safety precautions for the next 10,000 years? Fifteen thousand years ago the earth was covered in a sheath of ice, as Kai Erikson, a professor of sociology at Yale, observes. 'Ten thousand years! What vocabulary can we draw on to speak sensibly of such things? What compass can we use to find our way in such a vastness?' And how useful is the 10,000-year mark anyway? Ten thousand years is less than half the half-life of some radioactive elements. Does it help to look backwards in time? We know that the pyramids and the Sphinx of Giza are approximately 5,000 years old. Stonehenge is 3,500 years old and the Great Wall of China is 2,220 years old. If the menacing earthwork survives, what will people make of it? Will it resonate death and fear? What do we make of the Dead Sea Scrolls or hieroglyphic scripts? Will the markers be noticed or even noticeable? Will a site get flooded or buried as a result of a sudden climatic change? Will the monoliths and earthworks be destroyed by environmental pollution or war? Is our knowledge of geology and hydrology adequate to deal pre-emptively with a radioactive leak 600 metres underground at some time in the distant future?

Four days at the end of April 1995 cost the German nuclear industry 55 million Deutschmarks (approx. £22 million). That's how long it took to transport a single cask of spent fuel rods from a nuclear reactor in Philippsburg in south Germany to an interim waste dump at Gorleben in Lower Saxony. The convoy was unprecedented in German history. Seven thousand policemen armed with batons, dogs, water-cannons and canisters of mace battled with the anti-nuclear demonstrators on the road to Gorleben. Only a hundred protesters attempted to stop the cask leaving Philippsburg by train. By the time it arrived at Dannenberg, however, where it transferred to a lorry, thousands of protesters were ready and waiting. The cavalcade started its journey to Gorleben at 5.00 a.m. on 25 April, but the 35-kilometre drive ended up taking six and a half hours as protesters barricaded the road and chained themselves to the vehicle. 'It was a terrible day – we had nothing but our will to defend us,' said Marianne Fritzen, who helped organize the demo.

In a pine forest south of Gorleben looms a corrugated rectangle the size of a cathedral. It is a temple to modern technology, the most advanced nuclear waste storage facility in the world. Air cooled and computerized, the 120-metre-long warehouse is capable of holding 420 casks of high-level waste. Spent fuel is transported here from Germany's nineteen reactors. In the future, however, it will also store reprocessed wastes from Cap de la Hague in France and Sellafield in the UK. Each cask is locked into a separate monitoring system under the concrete floor.

'Waste is the bottleneck of the nuclear industry,' says Reinhard König, the managing director of Brennelementlager Gorleben (BLG), and therefore the person in charge of the site. Mr König admits that he was not expecting the scale of reaction he witnessed in April 1995. But he is undeterred. Indeed he is expecting another shipment from Cap de la Hague in three months' time and is convinced that opposition will fade as shipments continue. 'The protesters are not doing themselves any good. They are affecting tourism in the area. Holidaymakers are afraid to come here in case there is a demonstration.' Most of the anti-nuclear types are what König calls 'bohemians', people not originally from the area who are claiming social security, he believes. 'They're a tiny percentage of the community, but they have good connections with the media, so all one hears about is the protest.'

I went to Lower Saxony six months after 'Tag X', as that day later became known. Haydn's *Die Schöpfung* was being performed by the Kiel Philharmonic Orchestra at St Johannis Church in Dannenberg. The orchestra and singers had travelled south to play free of charge for two hours. The Protestant church is large but simply decorated, with only a few modern touches: a mobile in the style of Alexander Calder hangs above the pulpit, and the pillars are emphasized by brightly painted yellow and blue lines that meet in the vault. On the walls of the church are posters about famine relief bearing the legend 'Unsere Welt, Eine Welt' (our world, one world). By five o'clock on the day of the concert there were no empty seats in the church. Five hundred people – teenagers and pensioners, farmers and bankers, bohemians and conservatives – paid 15 Deutschmarks a head to listen to the music, which ended to a chorus of stamping feet and shouts for an encore. A stout woman in her seventies appeared on stage, handing flowers to the singers, and made a brief speech to bring the fundraising event to a close. As the church doors were flung open, and we shuffled out into the cold night air, I overheard another member of the audience say of the concert organizer, 'Isn't Marianne Fritzen a miracle woman? Did you know she started the whole thing?'

In 1973 Marianne Fritzen and twenty local farmers opposed a proposal for a nuclear power plant in Langendorf, 10 km from Gorleben. The proposal was defeated, but it resurfaced in 1976 in the guise of a 'nuclear park', an enormous complex consisting of two interim storage dumps, a reprocessing facility like THORP, and a permanent repository built 1,000 metres into the salt dome beneath the area. With the blessing of Ernst Albrecht, the Christian Democrat premier of Lower Saxony, the location for the 'park' was announced as being the pine forest a kilometre to the south of Gorleben, on the East–West German border. A few months later, 20,000 protesters stopped the traffic in Hannover, the capital of Lower Saxony, provoking a debate that focused on the reprocessing plant and on the risks attached to starting a plutonium economy. On 28 March 1979 140,000 people marched the hundred kilometres from Gorleben to Hannover in protest. It was a poignant coincidence that the demonstration took place on the same day as the accident at Three Mile Island. A month later Ernst Albrecht rejected the plans for the reprocessing plant, but in the heat of debate about reprocessing, the issue of waste disposal was more or less forgotten. In September 1979, drilling began for a permanent waste dump.

The Gorleben Action Group now has 800 members. 'We are not that popular here,' I was told by Miki Meunacher on the steps of the St Johannis church. 'In spite of what it may look like tonight. Of the 800 members, about a quarter come from Hamburg and Berlin. You see, one thing the nuclear industry has done to limit the opposition is pour money into the area.' No one is quite sure how much money, but estimates run to 350 million Deutschmarks. Before the nuclear industry arrived, says Lilo Wollny, another member of the action group, the village of Gorleben had an annual budget of 350,000 Deutschmarks. The Wendland was chosen, according to the group, because it is the so-called 'poorhouse of the Republic'. Unemployment in the area is currently running at 15 per cent. The area occupies a triangular annexe of land, south-west of Hamburg, that follows the River Elbe deep into the territory of the former East Germany. Before unification in 1990 the Wendland was surrounded on three sides by the Communist bloc, giving rise to a feeling of unease and vulnerability. There is no immediate access to an *autobahn* and there are limited rail links. 'This is the last corner of the FRG,' says Marianne Fritzen. 'The area is poor, unpopulated, isolated and we have a lot of land. In the eyes of the nuclear industry it's the perfect location. But I think what really frightens us is the feeling that if there was an accident at the waste site, it would be easy to section the area off and restrict entry.' Lilo Wollny says she is disgusted by the way the nuclear industry has ingratiated itself into the community. 'They arrived in their big Mercedes and the first thing they did was to make friends with the local councillors. They convinced them that once they got started, other industries would follow and we'd all be rich. They sponsored the local football clubs, paid for a community hall here, a swimming-pool there. They leased hunting grounds in the woods and gave lavish hunting parties. One had his daughter's wedding in Gorleben and invited the mayor and his cronies. They took them to Cap de la Hague to show off nuclear technology. Everybody got drunk. You have to remember that these councillors are used to making decisions about nothing more important than whether or not to repair a street or build a bridge. So for them, of course, it was all very impressive.'

The road to Gorleben takes you through the flat farmlands and woods of Lüneburg heath. It is impossible to lose your way, you just follow the graffiti. Every 50 metres a large radioactive symbol has been painted on the tarmac to warn visitors and to remind the locals. On yellow road signs, the symbol appears in black. In the woods the trees bear the legends, 'TAG X' or 'STOP CASTOR'. The

German anti-nuclear lobby takes pride in its work. Unlike the usual crude and hastily sprayed offerings you see elsewhere, these graffiti are stencilled on: the lines are crisp and exact. Or, at any rate, until you enter Gorleben when the graffiti suddenly disappear, and everything becomes quiet, as though you've reached ground zero. Here the tree-lined streets are neat and tidy. The gardens in front of the red-brick and timber farm workers' cottages look precise and trim. The village seems wealthy. It has two cafés, a restaurant, a shop and a nuclear information centre. It is hard to imagine, but this sleepy village in a sparsely populated corner of Germany has come to symbolize the country's anti-nuclear struggle.

Dr Rolf Meyer claims that he runs the two most important projects in Germany: the museum at Wustrow and the permanent waste repository at Gorleben. Dressed in a purple linen suit, garish tie and multi-coloured socks, he cuts an unlikely figure as a government spin doctor. His resemblance to Einstein and his strong liberal streak must be of concern to many in the nuclear industry. But Meyer has worked at the Gorleben Salt Dome Project since the beginning. He was offered the job in 1980 having completed a Ph.D. in hazardous waste management at Hamburg. It was one of the few jobs in the area on offer, and needing the money, he decided to take it.

Salt is an ideal material in which to isolate radioactive waste permanently. Salt is dry, for a start, and due to its plasticity, it is also stable; it doesn't crack or shatter. Also, salt 'creeps' – that is to say, it encases and seals anything you place inside it. In northern Germany there are over 200 known salt structures which originated 250 million years ago, at a time when the sea covering the German lowlands evaporated leaving huge quantities of minerals. Over millions of years, under the weight of strata, the salt deposits became deformed and began to rise as flat pillows, domes and walls of intensively folded layers of rock salt. The Gorleben waste repository will be constructed by sinking two shafts into the salt dome to a depth of a thousand metres. The shafts will then be connected by 120 km of galleries to contain the waste. The permanent repository, if licensed, will house all types of radioactive waste. The project is estimated to be completed in 2010 at the earliest, at a cost of 4.2 billion Deutschmarks.

I met Dr Meyer at the security gates outside the interim storage facility. The site, which is little more than two mine shafts and a number of administrative buildings in a clearing in the woods, resembles a high-security prison. Uniformed guards with Alsatian dogs patrol the inner wall night and day. Every 30 metres they pass a water-cannon mounted in a gun turret. The outer rim of the site is fenced off and layered with barbed wire. A mass of painted anti-nuclear slogans are daubed on the road approaching the entrance to the site. Every now and then someone flings a paint bomb at the wall. As you walk the perimeter, you are watched non-stop by the guards whose dogs growl and snap. Within minutes of meeting Dr Meyer at this fortress, he was telling me that he was not particularly in favour of nuclear power. This is an odd paradox. His job as public relations manager of a waste repository is to persuade the public that nuclear power has a bright future after all. 'I don't see my job as having to convince people that this is a fantastic project,' says Meyer. 'I want the public to contemplate the issues, to think about them. What we are doing here is pioneering work. How we eventually decide to use this place could affect us for an eternity. I feel a bit like the medical doctor in a war who is a pacifist, but has a commitment to help the wounded.'

Later, at a *Bierkeller* in Wustrow, his hometown, Dr Meyer explained that he did not regret taking the job, although he acknowledged it had its drawbacks. When he started work, his wife, Elke, a local teacher, was politely but firmly told by her colleagues and friends that her husband would not be welcome at parties or meetings. Dr Meyer's wife also recalls the time that they were thrown out of a local restaurant by its anti-nuclear owner. 'It has been very hard,' she says. 'I am cautious with people nowadays. But my husband is a good man. I think many of those who opposed him at first have now realized that he is unlike many of his colleagues in the nuclear business.' I asked her if she regretted Dr Meyer taking the job. After a long pause she said, 'I don't think that I should answer that question.'

But Rolf Meyer is troubled. He knows that if his efforts to 'sell' the repository to the public are to succeed, the company he works for, Deutsche Gesellschaft zum Bau und Betrieb von Endlagern für

Abfallstoffe (DBE), must adopt a less defiant attitude inside the community. The defiance is manifest in the battlements that surround the site. Another problem, he says, is a lack of philosophical debate. 'Everyone is getting on with their little job and the result is that the sum of the parts does not add up to much. There is this lack of thought about what we are doing.' The real problem with nuclear waste is its invisibility. 'How can we understand something we never see?' Meyer asks, staring gloomily into his beer. He acknowledges that the company has been badly affected by the protest. 'When a sparrow dies in the woods, we are immediately blamed as the culprit,' he says. 'So if a drop of brine is found in the shafts, the attitude is not to mention it to the press, because we know they'll kick up a huge fuss about it even if it turns out to be nothing. The workers feel that uncertainty and they don't know what to say and what not to say outside the plant. They are afraid and almost ashamed of the work they are doing. If you work for Porsche, everyone from the cleaner to the director takes pride in the company. There is no glamour in garbage.'

Public confidence in the German nuclear industry was severely damaged over what became known as the 'TN scandal', which came to light at the end of 1987. By the second week of the following year, two businessmen implicated in the press reports had committed suicide. Transnuklear (TN), a German company responsible for transporting four-fifths of nuclear waste in West Germany, was accused among other things of bribing Belgian officials to the tune of £7 million. In October 1986 a TN lorry overturned while returning treated waste to West Germany from a treatment facility at Mol in Belgium. The police investigating the accident discovered that the lorry's cargo did not correspond to a consignment note. Instead of low-level materials, it was actually carrying waste that had been mixed with radioactive cobalt and plutonium. In other words, its radioactivity levels far exceeded the regulations. It was discovered on further investigation that a total of 2,400 drums contained the cobalt and plutonium mix. Volker Hauff, of the German SPD, declared, 'We are faced with a moral catastrophe when not one company, but an entire network has displayed a considerable degree of criminal energy in violating regulations.' The affair resulted in a review of the 1959 German Atomic Law, to improve the oversight of the transport, licensing, storage and processing of nuclear waste. There was also a tightening of the West German export controls which disclosed other scandals regarding the sale of chemical weapons to Libya and of tritium to Pakistan and India. Many of the waste drums from the TN scandal are now stacked behind a concrete wall at the interim storage facility in Gorleben.

Lilo Wollny says that she was a latecomer to the struggle. She only got involved in 1978. During the Second World War, Wollny left Hamburg, her hometown, to stay with her grandmother in the village of Vietze, 10 km north of Gorleben. Fifty years on, she has yet to return home. 'It was my paradise. I could climb a tree and eat cherries whenever I wanted. The bakery was always full of cakes and sweets, and the dairy had fresh milk.' After the war, in spite of her desire to go to university, she married the baker's son and settled down to family life. It wasn't until 1977 that her unexceptional life was disrupted by proposals for a 'nuclear park'. Now in her sixties, Lilo is a short, robust, bespectacled woman with thick grey hair. Her conversation swings disarmingly from jolly raucous laughter one minute, to deadly serious hushed tones the next. As a young woman, she told me, she wanted to be a journalist. She has now made up for her unfulfilled ambition by virtue of her involvement in the anti-nuclear struggle. Indeed she represented the Green Party for four years at the Bundestag in Bonn. Her plain speeches, intelligence and common sense transformed her overnight from a provincial housewife into a very effective politician. 'I knew that if they put a huge industrial plant in my paradise it would change everything. It would change the whole structure of the place. People here were not very rich but neither were they excessively poor. We were uninfluenced by outside pressures and lived a simple life. So I knew that with the nuclear plant would come scientists, rich people and foreigners who would take over the council. We would become like niggers in our own homes.' Wollny began to read books about nuclear energy that her daughter sent from New York. She says she read everything, for and against. 'It was all very persuasive, and having read everything, I was just as confused. I thought who am I to decide? Then I read a book by Dr John Gofman. One sentence changed everything. It said that if 1 per cent of the radioactivity from a reprocessing plant escaped and there was no wind or rain for twenty-four hours, 8,000 square kilometres would be dead. Well I couldn't prove or disprove this, but I thought that I did not want to be responsible for something like that.'

Within months of her conversion to the anti-nuclear cause, Wollny had set up the Gorleben Fräulein and was organizing the march to Hannover. 'When I make a decision to do something,' she says, 'I become active, in the full sense of the word.' On 4 May 1980 Lilo Wollny, Marianne Fritzen and hundreds of local protesters occupied a clearing in the woods that was to be a test drilling site for the permanent repository. The demonstration became a *cause célèbre* throughout West Germany and brought Gorleben once again to national attention. The protesters declared the site as the 'Free Republic of Wendland' and within weeks, 3,000 anti-nuclear campaigners – men, women and children from all over Germany – were living in the woods in makeshift timber shacks, tepees and cabins. Trees were planted, gardens cultivated. There was a crèche, a kindergarten, a conference area for meetings and press announcements. As the media and politicians descended on the encampment, the police and the nuclear industry waited for the right moment to break it up. As protesters prepared for the inevitable battle, the Free Republic began to resemble a fortress with a perimeter fence, a parapet, an observation tower and a main gate. In the woods an underground network of connecting tunnels and trenches was built. After six weeks the police moved in wearing full riot gear. Dogs, horses, batons, bulldozers and helicopters were used to break up the demonstration in the course of a violent and bloody battle. 'I saw a man who only had one arm bent double on the ground while two policemen hit him with sticks,' Wollny remembers. 'I rushed over to stop them, but they started kicking me, punching me and pulling my hair. I remember shouting at one of them, "Why are you doing this? I could be your mother." He replied, "If you were my mother, I'd fucking kill you, pig." ' The Free Republic of Wendland was soon razed to the ground.

'The protest changed my life,' Wollny declares. 'I discovered what it was like to live as a free member of the community – it was incredible.' Nowadays she spends her time quietly cultivating her garden with her husband. Whenever there is a demonstration or fund-raising event, she makes sure to go along, but she admits that she doesn't recognize many of her fellow-campaigners any more. Nevertheless, she is just as enthusiastic about the cause and proud of the fact that it now includes three generations of Germans. 'I'm now a member of the Initiative 60 group,' she laughs. 'We stand between the police and the younger protesters. The police don't find it so easy to pull out grey hairs.'

When sun-rays light up the roads in the Gorleben woods, you can trace the history of the anti-nuclear struggle in the faded layers of graffiti that have washed away over the years. Every tree, wall, rock and bench in the picnic area in the wood opposite the waste site has been scratched, defaced or painted on. There are also the remains of campfires. A ripped tarpaulin hangs from a tree by a rope. But in spite of the prophecies of pro-nuclear hardliners that the battle for Gorleben will be over in a few years, the resolve of their opponents appears to be as strong as ever. After fifteen years in his job, Dr Rolf Meyer is understandably disheartened. The stress is beginning to show. Even his daughter Lena joined the demonstrations in Hannover and at Gorleben. 'What can I do?' he looks at me questioningly. 'Should I stop her from going. She is seventeen.' He wishes he could reconcile the two sides, if only for the sake of debate. 'This is a bigger moral issue than it is a technical one,' he says, 'and it is better to blah, blah, blah, than to boom, boom, boom. To be honest, I think that the final storage of nuclear waste is too big a problem, above ground or below it. I often wonder about grand projects like the Great Wall of China or the pyramids. How did they achieve something of that scale? Was it by using foreign slaves or military force or was it simply free will?'

'Dr Meyer is a jerk,' says Lilo Wollny. 'He claims to be on our side, he is also on theirs. But this fight has made my life worthwhile and I shall continue fighting until I die.' Lilo's husband was seventy-five last year. People they had not seen since the days of the Free Republic of Wendland turned up to celebrate his birthday. 'You see, we are family,' says Wollny. 'Those six weeks bound us together emotionally and intellectually. The spirit lives on in us.' In the meantime, the Gorleben Action Group will be preparing to obstruct the next shipment of reprocessed nuclear waste from Cap de la Hague.

11. *Vile Bodies and Quasimodo Fish*

In 1993 Taiwanese papers reported that fishermen on the east coast of Taiwan had been netting an increasing number of fish with large humps on their backs, and had nick-named the creatures 'Quasimodo' or 'Q' fish. A member of the *Therapon* family, the fish is similar in size and shape to the sprat which is eaten regularly in Taiwan. The 'Q' fish only grow to be 6 centimetres as opposed to the usual 10 centimetres. The fishermen had been netting the mutant fish in the vicinity of a nuclear power plant. Increasingly concerned that the strange phenomenon would place their livelihood in danger, they contacted Mr Fan Jeng-Tang, of the Taiwan Environmental Protection Union. Mr Jeng-Tang spent a day fishing off a pier next to the suspected power plant and caught several misshapen fish. As the only industrial complex in the area, the nuclear plant seemed to be the obvious polluter. The 'Quasimodo' fish immediately became a national media sensation. The Taiwan Power Company and the Atomic Energy Council entered the fray and began hotly to deny allegations that the deformities were caused by radioactivity. Samples of fish were tested and showed traces of naturally occurring radionuclides, but there were no traces of artificial contamination. Allegations of flawed findings followed and the debate continued. The authorities produced further evidence refuting any connection with the nuclear plant. At a press meeting Hsu Yi-Yun, chairman of the Atomic Energy Council, announced that the fish were safe to eat, whereupon he was presented with a jar full of live 'Quasimodo' fish for his own consumption. In the next few days a protest was staged by fishermen outside Yi-Yun's headquarters inviting him to a 'Q' fish dinner. The day after the protest, the Atomic Energy Council did a U-turn and released a statement saying that it could not rule out any cause of the deformity. Yi-Yun also told the press that he had never claimed that the fish was edible but if someone invited him to dinner then he would eat it. Meanwhile, the Taiwanese Environmental Protection Agency mustered a task force to look into the strange phenomenon. A group of scientists from the Atomic Energy Council, Taiwan Power and several universities, as well as two foreign scientists, continue to study the fish. One theory explored the *Therapon* species' sensitivity to temperature. Temperatures as high as 40°C have been recorded within 500 metres of the power plant, exceeding environmental regulations by 7°. One of the scientists raised some of the fish in a controlled environment of 38°C and found that the fish developed the same deformity. Records showed that the highest number of 'Q' fish were caught when the two reactors were operating at full power, causing the water temperature to rise considerably. When the second reactor was turned off, the temperature decreased to 32°C and fewer deformed fish were reported to have been caught. Blame for the high temperature levels was laid at the door of the Taiwan Power Company whose cooling system outflow appeared to have a design flaw. Instead of pumping the heated water straight out to sea, a pipe released it diagonally across the bay, ensuring that the high temperatures were confined to the 500-metre radius. The power company reported that it would have done something about the pipe earlier – it had known about the fault for many years – but claimed that the fishermen had objected, because the repair work would have hampered fishing.

The debate about the use of the world's oceans as a dumping ground came to a head in Britain in 1995, when the oil company Shell was prevented from sinking a 14,000-tonne oil platform, *Brent Spar*, in the North Atlantic. The plan, which had been approved by the Government, was to dump the rig in 2,000 metres of water between the Hebrides and Rockall at the North Fenni Ridge. It was reluctantly abandoned after Greenpeace launched a highly successful campaign involving high-speed chases in the North Atlantic and dare-devil helicopter landings on the rig. What cemented public opinion both in the UK and internationally was a desire to prevent big business from disposing of their unwanted technology in the most convenient and cheapest manner without a proper debate. It transpired that Shell had no detailed inventory of what substances the structure contained. The point that Greenpeace and its supporters made was that the deep-sea environment is relatively unexplored and is one of the most difficult ecosystems to study. Therefore it would be irresponsible to sink the rig if one could not accurately predict what would happen to the contents. It was also suggested that the sinking of the *Brent Spar* would set a precedent; it would open the floodgates to the oil and gas industries and condone the dumping of more hazardous toxic wastes. This argument touched the core of an emotive debate.

In the early 1970s the oceanographer Charles Hollister led an international study that looked at the viability of depositing nuclear wastes into the ocean. The project, which included scientists from Belgium, Japan, UK, France, West Germany, the US and Canada, was abandoned in 1986 because of adverse public opinion, political pressure and the efforts of environmental groups. 'I'm not saying that we should dump in the ocean,' says Hollister, 'but we should look at the possibility. Maybe not with a view to dumping now, but we might want to use the sea as a dump later. What I'm arguing for is that we should get the scientists on our side. We spent US$100 million looking into it and determined that there were distinct possibilities, but that there were still a lot of unanswered questions.' One thousand kilometres north of Hawaii lies a seabed to which Hollister refers as the most useless piece of real estate on the planet. There is no mineral wealth, little marine life and tests have proved that geologically it has been inert for 65 million years. The ocean is almost 5 km deep here and the bed is dark brown, of a fine-grained silt that has the consistency of peanut butter. Scientists have established that the silt is negatively charged and therefore attracts the positively charged ions of heavy metals like plutonium. As a result, the waste binds to the silt permanently. This would make it possible to sink canisters of waste using drill shafts, or drop torpedo-shaped containers from a ship. It is estimated that the waste would bury itself 30 metres into the mud. 'The deep oceans are very boring places geologically. Nothing has happened down there for millions of years. It is impossible to find that sort of stability on land,' argues Charles Hollister. 'The view that one must never interfere with the mother liquid (sic) of the planet because it is the last undamaged environment on earth, started with people like Jacques Cousteau and some of the early environmentalists,' he says.

'They didn't contribute much to ocean science, but they did a lot for romanticism. The sea has this sense of history, law and mystique about it, which has been the subject of children's stories and films for decades.'

The world's coastal waters resemble the dirt rim around a bath-tub. The waters off Hong Kong, Boston, New York and Vladivostock are all highly polluted. Since 1949, the UK has repeatedly dumped military radioactive wastes into the north-east Atlantic and off the Bay of Biscay. Between 1967 and 1977 the UK, along with eight other European countries including Sweden, Switzerland, Belgium, France and West Germany, dumped a total of 46,000 tonnes of radioactive waste in the Atlantic. In 1980 and 1981 the UK, one of only four countries continuing the practice, was responsible for almost all the alpha radioactivity and half the gamma and beta radioactivity dumped in the Atlantic. It is only in recent years that the cause and location of this waste have been discovered. In 1995 it was disclosed that the Beaufort Dyke, a 450-metre trench in the seabed just 15 km off the coast of Scotland not only contained millions of tonnes of unwanted explosives and chemical weapons but also 2,517 tonnes of radioactive waste. In 1981 a ship heading for the north-east Atlantic was forced to turn back due to bad weather and discharged its radioactive cargo into the Beaufort Dyke, out of convenience. The same year, 3 km west of the channel island of Alderney, 70,000 tonnes of nuclear waste was discovered dumped on the seabed at depths varying between 65 and 165 metres. The practice was outlawed in 1983. The United States stopped radioactive sea disposal in 1970 when it introduced the Marine Protection Research and Sanctuaries Act. The water around the Farallon Islands, 80 km away from San Francisco harbour, had been the Navy's dumping ground for decades. Some of the contaminated battleships and destroyers that were placed at ground zero in the first series of atomic tests in the Bikini Atoll were later dragged to Farallon and scuttled. In the early 1980s, Lt. Commander George Earl IV revealed how in 1947 he had flown secret missions to drop six large metal canisters into the seas off the New Jersey coast, where the radioactivity has since been measured as being 260,000 times greater than natural background radiation. Barges were also used to dump radioactive waste in the coastal waters a few kilometres off cities like New York and Boston.

'Seventy-one per cent of the world is covered by water and, despite all the dumping, 99 per cent of it is pretty clean,' argues Charles Hollister. 'You only have to travel 100 km from any land mass and you'll find the water's pretty clean. The idea that we abuse the sea at our peril is a fallacy that is put about by groups like Greenpeace.'

At 6 a.m. on a grey, wet day in early June 1987, Hans Guyt, the Netherlands-based international campaign director of Greenpeace, climbed out of his berth having been woken by a deafening noise. There was no mistaking that it was a helicopter: the *Sirius* was rocking from side to side. As he climbed on deck with the skipper, Willem Beekman, they could see the chopper hovering only metres from the boat. A package was thrown from the cockpit and it landed on the prow of the ship with a thud. They went to pick it up and as they did, the helicopter turned, dipped its nose and disappeared back towards shore. The parcel, delivered by British Nuclear Fuels Ltd (BNFL), contained three court injunctions: one for Greenpeace and one each for Guyt and Beekman. It warned them that if they continued their operation, they would be liable to court action. They withdrew to the warmth of the cabin to discuss it.

Hans Guyt and Willem Beekman sailed the Greenpeace vessel *Sirius* into the Irish Sea and dropped anchor off the coast of Cumbria opposite the Sellafield nuclear plant. Their intention was to block the pipeline that runs 2.5 km from the site into the sea. Guyt was an experienced campaigner who had fought for many years against allowing waste to be discharged in the sea. He had been to Russia in the early 1980s and had sent people to Murmansk to discover if the rumours about the dumping of the northern fleet were true. He had campaigned against the British ship *Gem* dropping waste in the Bay of Biscay and the Atlantic, and had campaigned against Sellafield ever since the 'Windscale Inquiry' in 1977 had signalled the go-ahead for the building of the THORP reprocessing facility. Both Guyt and Beekman felt that they had tried everything, save blocking the pipe, so they decided to continue despite the injunctions. Greenpeace UK had attempted unsuccessfully to block a waste pipe in 1983. They failed for a number of reasons: their equipment was not good enough and British Nuclear Fuels, suspecting some sort of action, built a metal cage around the end of the pipe. While the Greenpeace divers inspected the pipe, an unusually large and highly radioactive oily discharge was released into the sea. Workers at the site had accidentally diverted 4,500 curies of liquid material into the wrong tank where it had mixed with oily sludges. Since the error was irreversible the decision was taken to flush the mixture out to sea. When the Greenpeace divers returned to the surface in this radioactive slick, their Geiger counters went berserk. The news instantly became a scandal. The Department of Environment responded by warning the public not to use the beaches 25 km on either side of Sellafield, because contaminated seaweed and debris were being washed up. The warning was enforced for six months, after which Junior Energy Minister Giles Shaw went for a swim in a blaze of publicity in an attempt to calm public fears.

Sellafield is considered to be the biggest ocean-polluting nuclear installation in the world. The Irish Sea has been renamed by some the 'plutonium sea'. Radioactive discharge spreads up the Scottish coast, across the North Sea, and is carried by the Norwegian current to the Greenland Sea. It has even been detected in the Atlantic Ocean. Oceanographers can trace Arctic currents by following the spread of plutonium and caesium. Guyt says that Greenpeace's main concern was that the pollution was returning to the shore. 'The vast majority of things like plutonium accumulate in the seabed near the end of the pipe,' he explains. 'Over the years it gets washed towards the coastline, it is thrown by the spray and washed up on to the beaches where it dries and gets dispersed by the wind. This is the most likely pathway back to humans. Since the 1950s, nearly half a tonne of plutonium has been discharged from the pipe into the sea, and I feel that we have not heard the end of the health effects, by any means.' Piya Guneratne, a Sinhalese electrical engineer who started working at the nuclear site in 1955, recalls how surprised he and his wife were that very few people swam in the sea. After they had taken the odd dip, it was quietly explained to him by colleagues that people didn't like going in the sea because of the discharges. Piya refuses to eat the local fish to this day.

It took Hans Guyt and his team a whole day to find the pipe. The weather was bad, and the sea dark and muddy. They also had to contend with the added problem of the pipeline being flexible and moving with the currents. Using heat-seeking equipment, they eventually managed to locate the pipe and dropped anchor. The other members of the team were two professional divers, moonlighting from the Rotterdam police force. The most essential part of their equipment was a state-of-the-art underwater blowtorch, designed by a Dutch 'gentleman' burglar. The torch was capable of cutting a hole through any material.

The following morning, the sea was busy: five vessels, including a BNFL boat and the local coastguards, had gathered to find out what Greenpeace was up to. They sent divers down to investigate, and a battle, worthy of a James Bond film, ensued in the murky depths of the Irish Sea. While the BNFL divers attempted to stop them, the Dutch divers cut a 15-centimetre hole in the side of the pipe, inserted heavy-duty balloons and inflated them using compressed air. The effect was to completely block the pipe.

As soon as the job was finished, *Sirius* pulled anchor and headed for Ireland at top speed, leaving British Nuclear Fuels to unblock the pipe. Ireland was a critical player in the effort to stop the discharges, and Guyt was eager to persuade the Irish to take the lead in the Paris Convention, an initiative dealing with pollution from land-based sources. While publicizing their latest campaign in Dublin, Guyt and Beekman were notified that BNFL would be taking them to court. BNFL believed they had a good case against Greenpeace; by claiming damages and getting the court to confiscate the boat, they believed that they could effectively cripple the environmental group. What they hadn't banked on was that Guyt had set up an elaborate system of holding companies to protect their assets. The *Sirius*, for example, was owned by a small foundation in Holland. 'BNFL were very unhappy when they realized that they could not get at Greenpeace as an organization. We left them no choice but to go for me and the skipper!' says Guyt. The hearing was held at the High Courts in

London, and Guyt and Beekman were sentenced to three months' imprisonment for contempt of court. 'It was a strange court hearing,' says Guyt. 'At one stage BNFL tried to plead on our behalf and even appealed to the judge not to send us to jail. They were concerned that we would become martyrs.' In December 1987, Guyt began his stretch in Pentonville Prison in Central London. 'I was regarded as a special case because of the nature of my offence, and both the prison officers and the inmates were quite sympathetic. All the prisoners wanted to talk to me, and it took me at least a week to figure out that they weren't interested in what I had done, but in the blowtorch we had used!'

Because of prison overcrowding, Guyt was freed six and a half weeks later and returned to Holland to continue the campaign against Sellafield. 'It has always been cheap and politically convenient to dump at sea,' says Guyt. 'It is targeted because not a single voter lives there, but I regard the ocean as the common heritage of all mankind,' and he reminds me that that principle is enshrined in international law. 'The sea is not the private property of one country. The overriding question behind the London Dumping Convention is: How does it benefit others to allow countries like the UK to continue dumping? It's what is known as unequal distribution of cost and benefit.' Guyt feels that his years of campaigning reached a climax with the 1994 London Dumping Convention, where the seventy-four member states agreed to outlaw all dumping of any radioactive waste. All the states, that is, except one. Russia has not accepted the last amendment of the agreement, stating that they might have to continue the dumping of low-level radioactive waste.

It rained non-stop in Cardiff during January 1995. It was raining the afternoon Albert Stevens heard several urgent knocks on his front door. His neighbour Cathy Bewes had just seen the manhole cover at the back of her house burst open and a 1.5-metre geyser of effluent shoot into the air. Within minutes a flash flood of foul-smelling sewage had polluted five houses on Aberporth Road on the Gabalfa estate.

'By the time I rushed out the back it had come through the garden fence, up the steps of the back door and was seeping into the kitchen passage,' said Albert. 'I rang the council immediately to get some sandbags but they said that they were busy and I'd have to wait. By the time I got off the phone it was in the kitchen and heading into the hallway. I called the council again and they told me to take anything valuable upstairs.' The council eventually turned up, saying that there had been a lot of demand for sandbags that week, and dumped the ration of four in Albert's back garden. Unfortunately it was not enough to save the hall carpet from ruin. 'There was nothing I could do, really. It just kept on coming. I stopped it from going into the front room, mind,' he announced proudly.

Cathy Bewes and her husband Ian live at number 162 Aberporth Road. They were the first to see the sewer burst. The previous year a flood had destroyed the ground floor of their house, but they had built a wall in the garden in the meantime, so the damage was limited second time around. 'We'd been told by Welsh Water that it was a once in a hundred-year thing.' 'That's rubbish anyway,' interrupted Ian. 'It's happened three times already, since I've lived round here.' The Bewes's next-door neighbour, Bill Stamp, had also been flooded the year before. He threatened to sue Welsh Water until the company settled out of court. Bill's other neighbour was drunk at the time. Cathy couldn't get an answer from him when she knocked, so she assumed that no one was in. A short while later the neighbour's wife returned from work to discover the ground floor soaked in raw sewage and her husband asleep in the bedroom upstairs. The row that ensued was so fierce that the firemen were nervous of going in to help clear the mess up.

'The smell was horrible,' Albert Stevens's daughter Charmaine told me. 'It took over a week to dry out.' But in addition to the malodorous sewage the Stevenses discovered that something even more sinister had seeped into the house. As Albert was pulling up carpets over the weekend, he noticed that they seemed to be glowing. 'It was odd really,' he recalled. 'They were kind of speckled with lots of white bits. I mentioned it to Cathy next door, and she told me about Amersham International, the pharmaceuticals company, dumping waste into the sewers. It was the first I knew of it.' The subject of radioactive waste came up at the meeting in Gabalfa Community

Centre two weeks later. 'There was already a sense of outrage and anger about the flood,' Cathy Bewes explained. 'But when people learnt that the waste was radioactive they got very worried.' Many residents are concerned about the value of their property into which they have put all their savings. 'It's not easy selling a house full of radioactive sewage, is it?' said Cathy. At the Gabalfa meeting the complaints were heard by officials from Welsh Water Dwr Cymru and Cardiff City Council Engineering Department, who told residents that Welsh Water would not accept full liability. It declined to rule out the possibility of the same thing happening again, even though it proudly claims in its leaflets, 'As a responsible and caring company we do our best to prevent sewage flooding affecting an occupied property.'

The Gabalfa estate is one of the most popular council estates in Llandaff North, on the outskirts of Cardiff. An effective neighbourhood watch scheme keeps crime levels low. The wide residential streets are lined with trees and spacious pebble-dashed houses. Under the estate runs the Ystradyfodwg and Pontypridd trunk sewer on its way from the Rhondda valley through the alluvian flood plains of Cardiff Bay and into the Severn estuary. Built in 1891, the sewer is now urgently in need of repair. The original brickwork is cracked and has disintegrated in some places. The sewer leaks into the groundwater and its walls have burst a number of times under pressure from the increasing amounts of sewage that flow through it. The emergency gives rise to a wry scatalogical humour. 'Excuse my language,' I was told by local councillor Tony Earle, 'but every time a miner has a crap in the valleys it ends up flowing through my front room.'

Ever since 1979 Amersham International has lawfully discharged radioactive waste from its Cardiff plant directly into the sewer a kilometre or so north of Aberporth Road. A health science company, Amersham International manufacture (amongst other things) radiopharmaceuticals for use in nuclear medicine. These radioactive isotopes are used for imaging and scanning in an advanced sort of barium meal. After being administered to a patient, orally or by injection, the isotopes home in on a particular organ in the body. As a result, the consultant is able to examine the affected tissue by means of scanning the radioactivity. In the manufacture of these isotopes, hydrogen in water is supplemented by radioactive tritium (H3), and the carbon in carbon dioxide by radioactive carbon-14 (C14). Tritium has a half-life of twelve years. It has been used in fluorescent paints ever since it was discovered that radium, though more effective, was highly radioactive. It is also used in the trigger mechanisms of nuclear weapons because it enhances criticality. But the radiopharmaceutical process is very wasteful. Indeed the levels of tritium released into the Severn estuary by Amersham International are equivalent to those discharged from Hinckley Point B power station, which is generally considered to be the worst tritium polluter among civil reactors in the United Kingdom.

A former sheriff of Carmarthen, Tony Earle lives in a bungalow named Earle's Court – 'my wife's idea' – in Cardiff. Until his retirement Tony was the Labour councillor for Llandaff North. He still keeps his hand in campaigning for the community. In fact he is planning to sue Her Majesty's Inspectorate of Pollution (HMIP) for gross negligence. Three days after the flooding, HMIP and the Welsh Office renewed Amersham International's licence to discharge radioactive waste into the sewers. Earle is angry that not a single member of HMIP or the Welsh Office contacted the residents of Aberporth Road during those three days. Nor were the affected homes and gardens monitored for radioactivity, in spite of the fact that the flooding was well reported in the *South Wales Echo*. Gwilym Jones, the local Conservative MP, was also Parliamentary Undersecretary at the Welsh Office at the time. I asked the residents of Aberporth Road whether Mr Jones had offered his support. They responded with blank faces. One explained that if the MP ventured into Gabalfa nobody would talk to him. 'He's the wrong colour, if you get my meaning.' According to Tony Earle, the Conservatives' attitude shows a complete disregard for genuine public concern. 'The Tories didn't even bother to canvass this area at the last election,' he recalls angrily. 'Here is a threat of contamination. But they did not even take the commonsensical measures to protect public safety. They're a bunch of shits in my view. Excuse the pun.'

Discharges of tritium are frequently authorized in large quantities because the element is viewed by Her Majesty's Inspectorate of Pollution and the Ministry of Agriculture, Food and Fisheries (MAFF) as being relatively harmless. When concerns about public health were raised during the planning stages of the Amersham site, officials in Cardiff were told that health was not a planning matter and that any health concerns would be dealt with by government regulators. The regulators, HMIP and MAFF, do sometimes monitor the tritium levels in the environment but are relatively unconcerned. Tritium is a weak emitter of beta particles, which are unable to pass through anything thicker than a sheet of glass. It is therefore not seen as a threat to either public health or to the environment. According to HMIP, even if the sewage workers, a critical group, spent the maximum allowable 250 eight-hour days in the sewer they would still not absorb a dose in excess of natural background radiation. It is estimated that the residents of Aberporth Road would have to drink more than 30 litres of undiluted sewage – or 'liquor', as HMIP prefer to call it – to get an unacceptable dose by the standards of the National Radiation Protection Board. In other words, the very idea of a public health risk is viewed by the regulators as being preposterous.

Others believe that the regulators have done insufficient work on the subject, and might well have got it wrong. Hugh Richards, a town planner and environmental campaigner, is currently investigating the hazards of tritium in South Glamorgan. 'The inhalation of tritium as a vapour is more significant than the ingestion of tritiated water, because it is about 25,000 times more dangerous to the body than tritium gas,' he told me. According to Tim Deer-Jones, a marine biologist, the Severn Estuary and the Bristol Channel are the biggest receivers of tritium in Europe. Along the English coastline, nuclear reactors at Hinckley Point, Oldbury and Berkeley (now being decommissioned) all discharge tritium into the estuary, as do Amersham International and AWE Llanishen over the Welsh border. According to 1990 figures, tritium discharges on this stretch of the coast are ten times greater than those released from Sellafield. The estuary's currents move from Hinckley Point on the Somerset border towards Weston-super-Mare and across to Cardiff Bay. Deer-Jones argues that it is possible that significant amounts of tritium circulate in the air and water around Cardiff. He points out that nobody really knows the truth, however. 'I take a precautionary approach: if it is likely, then let us discover what that likelihood is.'

In other parts of the world, the tritium found in Aberporth Road would be regarded as a carcinogen. For example, there appears to be statistical evidence of an excess in childhood leukaemia mortality around two of Canada's nuclear reactors, Pickering A&B and Bruce A&B. These CANDU reactors are mainly tritium and carbon-14 emitters. A recent study found a 40 per cent excess in leukaemia mortality within a 25-km radius of the sites. Using a new environmental bill of rights, the citizens of Ontario are now pressing the Government to reduce levels of tritium in drinking water from 40,000 becquerels per litre to 20 becquerels per litre over the next five years. The Government has agreed to an objective of 7,000 becquerels per litre, but its target is still too high, says the environmental group Energy Probe, because it corresponds to about 700 fatal and non-fatal cancers per million people. 'There is a double standard in this country about what the authorities regard as a tolerable risk from non-radioactive substances and a tolerable risk from radioactive substances,' explained Norman Rubin, director of Energy Probe. 'Effectively our government thinks that we are more willing to get cancer from radiation than we are from non-radioactive substances.' In Rubin's view, this has to do with the nuclear industry's belief that its emissions are merely a subset of unavoidable, God-given, natural radiation. 'These guys believe that you can't talk about radioactive pollution without talking about airplane journeys, sleeping with your wife or radon in basements,' he adds. 'We're supposed to be comforted by the concept of ALARA (as low as reasonably achievable) for radioactive pollution, which, to speak crudely, is a crock of shit. For example, there's a nuclear station in New Brunswick that discharges tritium into the ocean and not into a drinking-water pathway. It is therefore allowed to discharge significantly more than it would have done if the discharges were connected to the drinking-water pathway. This is considered as low as reasonably achievable. It's a joke.' Welsh Water says that it is now taking measures to prevent more flooding in Gabalfa. It has also applied to the National Rivers Authority (NRA) to divert the sewage into the Whitchurch Brook, which leads into the River Taff. But this will take it through the new Cardiff Bay Development, which some argue would contaminate a much greater area.

In the meantime, Albert Stevens has just finished redecorating the house. Not being insured for its contents, he made a claim to the water board and has just received payment. He hopes Welsh Water will build a wall strong enough to resist any future flooding, he told me as I looked over the fence at the back of his garden. Six months after the flood, the foundations of a wall had been laid, and there were odd signs of activity, including a bag of cement and a few breeze blocks. Nevertheless, the wall did not appear to be high on the priority list. The plot beyond the wall is scrubland. Children use it regularly as a shortcut to the shops, riding their bikes over the sewer. Sometimes they use the manhole covers as ramps from which to launch the bikes into the air. Ultimately there is a plan for the land to be made into a playground. 'But what we want is a survey of the soil,' complains Albert. 'You see, the land has been flooded so many times now, we just want to know how contaminated it is.' In 1995 a request by Amersham International to increase its tritium discharges was under consideration.

'On Wednesday, 28 March 1979, at 6.00 a.m., my husband and I were outdoors. We had a "clean" metallic taste in our mouths. My son and I were outdoors from 7.45 a.m. to 10.00 a.m. Later that day we both had sunburn effects on our hands and faces. Thursday we drove to the west shore, to Ashcombe Veg Farm near Grantham. During that drive my eyes were watering and burning. It was so bright, it hurt to see. Friday morning, after hearing sirens, church bells and the radio news of uncontrolled releases, we evacuated. Tuesday evening, my husband and I returned home for winter clothing, medicine and teddy bears. During our brief two-hour trip home I encountered an "unusual event". The problem I observed was the accelerated growth of my umbrella plant. New growth measuring more than a 3-inch by 5-inch card had appeared within five days! We returned home eight days later. Sometime later (I don't remember how many days), while giving my two-year-old son a bath, I noticed a small wad of hair in the tub. His hair was thin, you could see his scalp.

'In May 1979 my daughter picked a bunch of wild field daisies, with two grossly deformed flowers among them. I also found three dandelions in my backyard that appeared to be similarly deformed. I have found many of these every year since 1979 (my neighbour who lived here over twenty-five years had never observed this before). To date I have found plant abnormalities in these areas: Londonderry township, Derry township, Fairview township, Harrisburg, Newberry township, Lower Swatara, Swatara and Upper Allen township. I can't say "all" abnormalities found were caused by radiation or chemicals from the Three Mile Island accident, but I believe the fallout from the accident has caused most of the effects I've seen.' Mary Osborn of Swatara township, Pennsylvania, 14 January 1985 (Mary Osborn concluded her document with pencil sketches of the growths she had witnessed on her plants.)

Cornelia Hesse-Honegger's most recent paintings are stacked neatly against the wall of her studio in Zurich. Although she has had regular exhibitions of her work throughout Europe, she does not sell any of it. As a matter of fact, she has never tried to. 'People don't want pictures of such things,' she says modestly. Her paintings depict the world of the common or garden insect. The abdomen of a bug is enlarged to become a giant bloated belly, measuring, in some cases, a metre. The thorax of a beetle is overwhelming; the giant eyes of a housefly gloat at the viewer. Her insects are *insects* that defy the onlooker. Billowing layers of shiny multi-coloured skins, deep reds and yellows, rich browns and blacks, spotted and bubbled symmetric shapes expose themselves on the sterile, white paper. On closer inspection, however, each painting has a darker side to it. Rather than the malevolent monster organisms of Hollywood B movies, her subjects are innocent victims. These insects are unnatural, abnormal, deformed, mutated, distorted. At the bottom of each painting lies a clue: '*Found near Sellafield 1989*' . . . '*Found near Chernobyl 1990*' . . . '*Found near Three Mile Island 1991*'. Hesse-Honegger's work confronts us with a view of our world that we would prefer not to see. We are compelled to witness our own destructiveness, our failings and weaknesses through the subjects we have poisoned and destroyed. 'I think of my work like a crime novel,' she says. 'I am trying to discover who the murderer is, and every now and again I discover a clue, but mostly I find bodies.' Her most recent works have just returned from an exhibition in Hamburg. The paintings are of insects she collected in the towns of Obermarschacht and Niedermarschacht, near the Krummel nuclear plant in Lower Saxony, Germany. The insects have uneven and misshapen wings, others have blisters on the thorax and deformed legs. The Krummel plant has been at the centre of a fierce public and scientific debate for the last five years. The number of children with leukaemia in the towns surrounding the plant is much higher than the national average.

On the easel in her studio is a painting in progress. The form taking shape is that of a *Heteroptera* or leaf bug. She has studied the leaf bug for much of her career and has become highly knowledgeable of the subtleties that distinguished not only one bug from another, but one antenna from another. The precision and detail of her paintings are breathtaking. To the left of the easel is a jar full of tiny brushes. Each brush has been custom-made. A cluster of hairs on one brush forms a fine point, another has been cut off abruptly to form a blunt end, others have been plucked so that only three or four hairs remain. A large microscope and a tupperware sandwich box containing the dead bugs lie next to the jar. The bugs are held in place by two pins and are arranged in rows, twenty across and ten deep. Under her microscope is a member of the *Heteroptera.* It is an attractive insect with a thick, square, shielded body that averages 6 millimetres in length. It has long striated antennae and an outer armour of golden browns with black dots and the occasional hint of green. A family of *Heteroptera* will pick a single succulent plant and make it their 'home' for many generations. It is Cornelia Hesse-Honegger's favourite insect for this very reason, as it enables her to return to a plant year after year with the guarantee of finding the same family. This provides a fairly accurate assessment of the amount of radioactivity that has been absorbed. *Heteroptera* feed directly from the fluids in the plant and are therefore one of the first links in the contamination chain to be affected. Radionuclides wash into the soil or settle on the leaf, where they are absorbed by the plant and, in turn, digested by the leaf bug.

Cornelia Hesse-Honegger started her training as a scientific illustrator in 1961 at the Zoological Museum of Zurich University. For much of her career she has been employed by geneticists, painting fruit flies (*Drosophila*) at the Zoological Institute. Her skills have been used to document accurately, and in minute detail, the noticeable changes that occur in fruit flies when they have been given chemicals in their food. The fruit fly has played an important part in the field of genetics. The fly breeds every three weeks, allowing scientists to identify genetic changes over many generations in a single year. The *Drosophila* became famous in the 1930s when the Nobel Prize winner Hermann Muller discovered that radiation caused genetic mutations in the fly. In his Nobel Prize acceptance speech he told the audience, 'Good mutations are so rare that we can consider them all as bad.' At the time of Muller's findings the more established professions of medical science and physics assumed that standards for an acceptable radiation dose could be based upon the observable effects on the body. The work of Muller and other biologists challenged the belief that there was a threshold, or level of tolerance, for radiation. They argued for the cumulative and long-term risks of radiation to be considered and proved that doses below the set tolerance level affected individual cells. This thesis was only seriously discussed by the International Commission of Radiological Protection in the 1950s and resulted in a tightening of radiation standards.

Cornelia Hesse-Honegger has attempted to combine in her career her role as a scientific illustrator with her impulse and energy as an artist. 'I never wanted to be an artist,' she says. 'My parents were both artists and knew a lot of famous painters, and I had no longing to be like them. But when I worked at the institute I quite naturally coloured the backgrounds to some of my work. The scientists didn't like this at all; it wasn't scientific!' In the early 1980s she realized that this balance was being upset; she was becoming absorbed into the world of scientific experiments and anaesthetized to the deformities she was painting. 'My life at the time was so comfortable. I spent my days with my family, baking bread and painting flies. Yet all around me atomic bombs and chemicals were being tested. I felt that I needed to express my political feelings through my art. I became preoccupied with the idea that some things in our natural surroundings have changed in the same way as the lab flies had. I had always relied on the knowledge of the biologists to explain how a plant or animal should look; they, I thought, knew about nature. There were individual differences within the species, from one animal to the next, but one could always find out to which species something belonged simply by consulting books. Every creature, large or small, is determined by evolution and genetic information. That is what defines the appearance and beauty of our natural environment. I no longer felt that this was true. I began to question what was normal. Were the leaves and bugs around us normal or a mutation? Perhaps we are all living in a world that has mutated over the centuries. I became highly sensitive to everything,' Hesse-Honegger explained.

By the mid-1980s she was working more and more on her own studies, painting the flies, bugs and plant life near her home. From her observations she discovered that many of the species were disappearing. What had been common, now no longer existed. It was undetectable to the naked eye, but plain as day under the microscope. The world was changing in minute detail. Pigments were changing, forms were altering. 'These changes are being allowed to occur unnoticed,' Hesse-Honegger continued. 'No one is documenting them. We cling to images of our world through books, but we do not really look at what is around us. Our idea of what the world is like no longer corresponds to the reality.' Then in April 1986, she heard the news about Chernobyl.

Cornelia Hesse-Honegger describes the impact of the Chernobyl accident on her life as though it were a 'Before Christ and After Christ' experience. The year before the accident she had been given irradiated fruit flies to paint. She was shocked by what she saw under the microscope. 'They were the worst deformities I had ever seen. Some had legs growing out of the antennae, others had growths on their eyes, colours varied and their wings were misshapen and curved. After the accident at Chernobyl, it was clear that the sterile, controlled environment of the scientific laboratory experiment had become irrelevant. The experiments were now occurring in reality.' Although the scientists at the institute told her that the levels of radiation from Chernobyl would be too low to affect the natural surroundings, Hesse-Honegger left for Sweden, the first country to register the fallout. 'I took my paints, specimen jars and microscope to the parts that had received the highest fallout, Gavle and Oesterfarnebo,' she recalled. 'I set up my studio in the hotel room and looked for *Heteroptera.* I saw deformations on their antennae, legs and wings. Many had asymmetrical bodies, others formed an "S" shape. I saw leaves that had peculiar shapes, clovers that were dark red with yellow flowers instead of pink flowers. I heard from the local vet that there had been a number of calves born with deformities. The local people were terrified and very depressed. They didn't know what was safe to eat and what wasn't. They had been told that they could eat everything from their gardens except berries and mushrooms, but everything was so uncertain that nobody knew what to believe. I got in touch with the local scientists at the University of Umea to find out if they were undertaking similar studies. But no one showed any interest in my work, nor were they doing their own.' Hesse-Honegger returned to Switzerland with her findings. Excited and appalled by what she had found, she immediately notified the institute expecting them to be fascinated. Once again she met with disinterest. The attitude of the entomologists and radiobiologists was that her work was mere assertion; science was objective, art was subjective. Her work had no place in the scientific arena. One scientist working at the Institute of Radiation Study in Zurich was interested enough to suggest placing her specimens on photographic paper and leaving them for a month to see if the radiation dose was large enough to affect the paper. But the project was cancelled because his superior would not allow him to get involved. Another scientist provided her with maps of the fallout from Chernobyl and data referring to the wind conditions and rainfall at the time. However, he made it clear to Hesse-Honegger that she should never call him at work. 'Scientists in Switzerland are afraid and react emotionally and aggressively when challenged,' she told me. 'One referred to me as a bitch, another would show his students my work to explain the extent to which some stupid people will go to prove that low-level radiation is harmful. Later, when I published my work, a scientist said I was personally responsible for the fact that a group of women in southern Switzerland had had abortions after Chernobyl.

'I now realize that it is totally logical that a woman artist should be doing this work. Switzerland is a very patriarchal society: women didn't even get the vote here until the 1970s. As a woman, I think that I have a greater empathy with my surroundings. I have much more freedom than a scientist. An artist is not fixed by some doctrine or philosophy; I can quite easily say that I do not believe what I believed yesterday. Artists are emotional and subjective. What we do has nothing to do with logic or rationality. We work with random systems. The artist does not need to know what he is doing, the scientist will pick that up later and explain it. But the initial exploration is always done by the artist. This has always been the case. In the fifteenth century art was a hundred years ahead of science. Nowadays that gap has almost completely closed, but I truly believe that we cannot really see something that has not been painted or put into an artistic form. It simply does not exist until then. I believe that the artist should be incorporated into the academic world, integrated into the learning of every subject.'

Undeterred by the response from the scientists, Cornelia Hesse-Honegger continued her research into the fallout cloud from Chernobyl. This time she went to Ticino in southern Switzerland. Once again, even though the levels of radiation were lower than in Sweden, she found similar deformities. She collected fruit flies from the area and reared them in her studio using techniques she'd observed at the institute. At the end of 1988 she published her findings in the national magazine *Tages-Anzeiger Magazin*. 'The fruit fly is like a holy cow to the geneticist,' she wrote. 'I have not only reared my own, but have painted and observed the mutations over several generations. This was sacrilege for them.' There was total silence from the institute; not a single article appeared in response. However, the response from the public was astounding. 'People sent me deformed plants that they had found on the German border. Others told me of the deformed insects they had seen. I was sent a kitten that had been born with six legs and two tails and an open stomach. One woman, I remember, called me at the studio to say that her daughter was going to Sweden; she wanted to know if it was safe.' Hesse-Honegger's work had quite suddenly struck a chord with the public. They could clearly identify and recognize the effects without the scientific explanations. 'I realized that I could change the way people looked at their environment,' she explained. 'This was a great comfort to me, I knew that I didn't need to be embraced by the scientist.'

In 1989 she began to study the countryside near nuclear power stations. She explored the Melano and Mendrisio Leibstadt areas of southern Switzerland. Over the next few years she travelled to Sellafield in England, Creys-Malville in France, Chernobyl, Three Mile Island and Peach Bottom reactor in the USA. In each place she found further deformities. Ulcers on the antennae and genitals, malformations of the wings, unusual brown markings on the chitin. 'I can now say that certain deformities can only be found downwind of a nuclear power plant,' she said confidently. 'I discovered that after one Swiss reactor had been closed for a year and a half the numbers of deformities I found decreased dramatically. 'What seems to happen is that the symmetry of the animal is disturbed. The malformations always occur on only one side of the body, and they can be tiny disturbances. One cell carrying information, for example, for the growth of the left wing is damaged resulting in a deformity, while the cell carrying information for the right wing is perfectly normal. These are not generational mutations – the effects are much more chaotic than that. So one year a generation of bugs can be affected, and the next year another generation appears normal.'

When it finally came, the response from the scientific establishment in Switzerland to Hesse-Honegger's work was straightforward. It argued that her findings could not be the result of man-made radiation, because the levels recorded were simply not high enough. The scientists had two main arguments. Firstly, their work in the laboratories showed that that type of damage to an insect results from much higher levels of radiation. Secondly, radiation levels from the accident at Chernobyl and the emissions from a nuclear power plant are insignificant when compared to those of background radiation. These arguments send Hesse-Honegger into a rage. 'The way of thinking at these universities and institutes,' she says, 'is old fashioned and too linear. Radiation is not logical, it works in a very random way. It is like shooting a gun. The number of bullets in a gun only affects the likelihood of being hit, but if there is only one bullet in the gun and you are unlucky enough to get hit, you suffer the same consequences. Radiation works in the same way: even if you only ingest one isotope, you can still get cancer. The radiation experiments conducted on fruit flies are done using X-rays,' she explained. 'There is a big difference between X-rays and the rays from artificial radiation. X-rays travel as a wave, whereas the rays from plutonium, for example, are stochastic. To compare the two is simply unscientific.' She regards it as equally unscientific to compare background radiation with man-made radiation. 'It is like comparing a potato with a car. We are talking about Chernobyl!'

Cornelia Hesse-Honegger has never met Mary Osborn of Lower Swatara near Three Mile Island but she says that in nearly fifteen years of collecting bugs, she is the only person that she knows of who has done similar studies to her own. In all that time there has been only one scientist who has been supportive – an entomologist in Sweden who collected *Drosophila* in the same part of the country as her. They would swap samples and notes with each other, but rather than openly meet to discuss the work at his university, he would insist that they meet in clandestine fashion on street corners and in cafés off campus. She is shocked and saddened that scientists do not get out of the laboratory to study the environment in the way she does. 'They are trying to protect radiation from the people, rather than the other way around,' she said. 'I think they are working for the industry. They must be aware of the disturbances to nature, they are not stupid. But if they have done their own research, they have kept their findings secret. They know enough about radiation to start telling the people the truth. But there is a career structure and hierarchy in the scientific world that manipulates and determines the outcome of so much of the work they do. I remember when I was working at the institute we used to manipulate what I saw by touching things up slightly or by altering a colour. Once I was asked to make the hairs on a *Drosophila* a little longer so that it would conform more to their idea of what it should look like. It was acceptable to me then because I believed in what they were doing.'

In 1994 a survey of the Hanford nuclear reservation's insects, birds and plants discovered a new species of bladderpod plant, three new species of bees and four new species of leafhopper insects. Continuing the research, the Nature Conservancy of Washington State discovered a type of buckwheat, a member of the *Eriogonum* family, that had never been seen before. The windswept plant was found growing close to the earth with small bunched blue-green leaves and 5-centimetre stalks with yellow flowers. The study was prematurely discontinued in 1995, after its budget was cut.

12. *The Waste Opportunists*

In November 1995, the Chechen rebel Shamil Basayev disclosed on the Russian television station NTV that he had placed a 'hot' package somewhere in the centre of Moscow. The Russian authorities, who had long feared such a reprisal, immediately deployed the army systematically to scour the city with mobile radiation sensors. At last the soldiers found a package of caesium-137 lying covered by a thin layer of snow in Izmailova Park in east Moscow. The official press release stated that its radiation levels were only thirty times natural background radiation and therefore not a hazard to the public. The stunt was out of character. Six months earlier Basayev had taken a thousand hostages at a hospital in the southern Russian town of Budyonnovsk. A protégé of Chechen rebel leader Dzhokhar Dudayev, he has since warned that his next target will be a nuclear power station.

Up to now the world has been lucky to escape the wrath of the nuclear terrorist. Yet it was not only in Russia that Basayev's action unlocked a deep-rooted fear. Having witnessed the Oklahoma bombing and the sarin-gas attacks on the Tokyo subway, nobody can underestimate the hypothetical danger posed by a political activist or religious fanatic or madman obtaining a nuclear device or perhaps highly radioactive material. The rudiments of atom-bomb-making have been widely known for decades. Indeed any physics graduate, with the right ingredients and technology, would be able to assemble a crude but lethal prototype. In the 1960s a book was even published about how to build the bomb. The problem, however, is getting hold of the necessary fissile materials. And for this reason the nuclear industry continues to be the most regulated and security-conscious industry in the world. Nevertheless, the success or failure of any regulatory system depends upon its safeguards, its checks and balances, being properly maintained. In recent years the nuclear equilibrium has been upset by miscellaneous arms reduction treaties and hectic decommissioning, not to mention economic and political chaos following the break-up of the former Soviet Union. Keeping tabs on fissile material nowadays is proving to be difficult. Since 1990, various quantities of plutonium, enriched uranium and the now infamous red mercury have been discovered in transit by the German police. The first seizure occurred in May 1994 at Tengen, south Germany, when police raided the garage of a businessman and found 5.6 grams of highly purified plutonium-239 stashed in a drainpipe. In June of the same year another German swoop uncovered 0.8 grams of enriched uranium. Four Slovaks, one Czech and a Bavarian property dealer were arrested. Two months later at Munich airport, a suitcase belonging to a Colombian businessman was found to contain 363 grams of plutonium-239. The cache represented the first of 4 kilograms of plutonium that was being sold for US $250 million. In December 1994 a further 2.7 kilograms of enriched uranium were discovered on the back seat of a car in Prague. Nobody knew where the fissile contraband belonged. Indeed one conspiracy theory even accuses German Chancellor Helmut Kohl of orchestrating the snatches in order to curry political favour. But in spite of the vigorous denials from the Ministry of Atomic Affairs (Minatom) in Moscow, the finger points to Russia, although now it is feared that Ukraine has superseded Germany as the channel favoured by nuclear gangsters smuggling material to countries such as Iran, Iraq and Pakistan. In September 1995, while investigating the murder of a journalist, Ukrainian police discovered three jamjars containing enriched uranium pellets in a flat in Kiev. At the time, the flat was being rented by a Russian army officer. The uranium, it was finally discovered, had come from a reactor in Moscow. According to the city's Federal Department for the Protection of Strategic Objects, there have been over 900 attempted breaches of security at Russian nuclear facilities since 1990. The department claims that each attempt has been thwarted. Vladimir Kuznetsov, former chief of the Nuclear Safety Inspectorate in Moscow, told me that the lack of security in up to fifty nuclear installations around Moscow was a matter of grave concern. The security fence around the Kurchatov Institute, Russia's premier nuclear research centre and perhaps the origin of the Tengen seizure, is equipped with a system designed as long ago as 1953. The guards on the gate do not have the resources or the technical back-up to make sure that people leaving the institute are not carrying radioactive materials. And the further away from Moscow you go, Kuznetsov explained, the more acute the problems of security and inadequate facilities become. He cited both Arsamas 16, where nuclear weapons are now dismantled, and Chelyabinsk as being insecure locations.

Knowledge is an obvious means of security: you have to know exactly what it is that you are guarding. In other words, a rigorous accounting procedure is a fundamental principle. Over the years more than fifty attempts have been made to extort money from American businesses or the Government using the threat of nuclear terrorism. To date all have been ignored, without consequence, on the basis that the official documents show that no fissile material is missing. It has been suggested, however, that many companies maintain an unaccounted slush fund of plutonium in order to balance the books whenever the plutonium accounts are inexplicably in debt.

The end of the Cold War has given rise to many other waste opportunities. To coin a phrase, 'Where there's radioactive muck, there's brass' has never rung more true. In the last five years the cost of the clean-up, of decontaminating and decommissioning the United States' seventeen bomb production sites has risen to a trillion dollars. The nuclear waste industry has a bright future. The Department of Energy's annual budget, although recently cut by Congress, still hovers at the 6-billion-a-year mark. Unsurprisingly, many international firms have their sights firmly set on this nuclear honey pot. British Nuclear Fuels Ltd (BNFL), for instance, now adapts its waste technology to suit the American market, and as a result is one of the most successful foreign businesses in the market. 'The Americans Always Want the Best, so Naturally they Chose to Work with Us' ran the smug copyline in a recent advertising campaign. 'We intend to be on an equal footing with the parent company in the not so distant future,' I was told by James Schlesinger a spokesman for BNFL Inc., the US subsidiary, 'A future where we exchange and utilize our technologies freely.' The same advert referred to a five-year, US$3.5-billion contract at the Rocky Flats site where BNFL has formed a partnership with the American giants, ICF Kaiser, CH2M Hill, Westinghouse and Babcock & Wilcox. Eastern Europe is also providing a variety of openings. During the Cold War, most of the nuclear waste from Warsaw Pact countries was sent to the Soviet Union, which had a monopoly on uranium enrichment and processing. Nowadays those services are routinely put out for tender, with contracts often going elsewhere in Eastern Europe, and Western firms now competing unashamedly in this expanding new market.

Inevitably, the nuclear waste market in America has provided openings not only for entrepreneurs, but also for criminals. Ray Peery, a forty-year-old executive director of the Central Interstate Low Level Radioactive Waste Compact Commission, was arrested in 1991 for embezzlement. The commission was originally set up by five American states to find a suitable location for low-level waste. It paid funds into an account held at a bank in Lincoln, Nebraska, where Peery was accused of transferring US$600,000 from the fund into his own name and charging the amount to legal fees, office supplies and other expenses. Nevertheless, his various misdemeanours pale in comparison to the life and work of salesman extraordinaire Fred Bierle. Bierle (pronounced 'buy early') is responsible for the two largest waste sites in America today, Richland, Washington State, and Barnwell in South Carolina. His success at siting a dump is apparently due to a mixture of charm and unaccustomed zeal, both of which are attributes that he has displayed consistently throughout his career. In the past his entrepreneurial exploits included marketing a pick-up truck that ran on a combination of hay, wood and weeds, and patenting a machine to convert cherry pits, corn stalks, rubber tyres and chicken manure into synthetic gas. 'One part of you says you ought to know better about some of the things he tells you,' a former colleague observes. Bierle's career in nuclear waste began in 1962 when the Atomic Energy Commission, the predecessor to the Department of Energy, outlawed the dumping of commercial waste on military sites, thus heralding opportunities for the private sector. The unenviable task of designating a nuclear waste site was thus conveniently taken away from the central government and handed to venture capitalists like Fred Bierle. Immediately Bierle formed a partnership by the name of California

Nuclear Inc. with a nuclear engineer and a health physicist. Within a couple of years the company had licensed its first site, Richland Burial Facility, but after pausing to congratulate himself on a spectacular debut, Bierle went in search of the next conquest. He picked the small town of Sheffield, Illinois. Bierle moved lock, stock and barrel to Sheffield and began to insinuate himself and his family into the community, all the time extolling the economic benefits of living next door to a waste site. He assured Sheffield that the waste was harmless. 'The material we handle is sweeping compound, glassware, rags, clothing, contaminated tools and even chairs,' he told locals. 'In some cases the rubbish we put in drums and bury is no more radioactive than an alarm clock.' He also promised to be at the site every day and would therefore receive a higher dose of radiation than anyone else. Within a year and a half, Bierle had not only convinced the town of its suitability, but had also persuaded the regulatory boards that in terms of hydrology and geology Sheffield was ideal for nuclear waste. As the waste trucks began to roll into town, Bierle packed up and left, having sold the company to the Nuclear Engineering Company. Today, the Sheffield site has been closed down owing to a leak of radioactive material from the site. Besides clothing, glassware and rags, the rubbish buried there included parts of a reactor, 13.5 kilograms of plutonium and 31.5 kilograms of enriched uranium.

Ron Fino is the son of the former boss of the Buffalo Mafia. He became an informant for the FBI in the early 1970s and continued to pass information about the workings of the Family until his position became untenable. One day he just disappeared. Since then Ron Fino has been living under constant guard and a false identity as part of the Government's witness protection programme. In return, he is obliged to testify for the prosecution at various court hearings about Mafia scams. Fino was never a 'made' guy – that is to say, he was never formally inducted – but he was privy to the Mafia's operations on account of his family connections.

Hidden in the pine forests an hour's drive south of Washington DC, Quantico is the FBI training centre for special agents. On arrival, I was escorted into a large empty classroom and introduced to Fino. Short and stocky with a thick neck and dyed black hair, he resembles the classic 'wise guy' of Hollywood legend. His street talk is fast and tough but also articulate and intelligent. As Fino outlined Mafia tactics for me, our two chaperons, FBI agents, sat against the wall looking bored. I was told how organized crime controls the labour unions, how it extorts money through pension and insurance fraud or stock manipulation, how a Mafia family can destroy a competitor by flooding the company with money, thus inflating the share price, before suddenly pulling out and ruining the company. 'The mob makes money on anything and everything,' Fino explained. 'Nothing is outside their sphere of influence. If they want something, they'll get it. It's as simple as that. The Mafia are good at manipulating the freedom of American society, the best thing that the Government can do is create a new law or regulation. The Mafia love regulations because they can manipulate the situation. Organized crime is the natural result of a capitalist society: it thrives on a supply and demand economy. At the moment the big thing is the environment. The Mafia are big in the toxic waste disposal business, y'know, illegal dumping, scrap, hazardous waste being shipped abroad, that sort of thing.' Although most of the activities he described concerned chemical wastes, it cannot be long, if they are not doing so already, before organized crime gets involved in the nuclear market. With more and more regulations being introduced each year, the cost of disposing, storing and transporting nuclear waste has become so prohibitive that many firms are eager to pull out of the industry. The situation is ripe for exploitation. John Weingart, director of the Waste Siting Board in New Jersey, a state frequently linked with the Mafia ever since the 1930s, refers to the Draconian regulations that protect nuclear enterprises from *La Cosa Nostra*. 'The history of all companies used are thoroughly checked,' he told me, 'and all the company stock-holders are finger-printed.

According to Ron Fino 'the Mafia's involvement in any shakedown is completely hidden.' The money is laundered again and again until it's clean and nobody's the wiser.' The same thing happens in public life, he says. 'The Mafia invests in a politician using a puppet. I could name many politicians who are in bed with the Mafia and don't even know it.' He adds that the powerful and wealthy Russian Mafia is now involved in the smuggling of fissile material. Its recent activities in the United States, particularly in New York, have posed a significant threat to the American Mafia, which is rumoured to have agreed a system of mutual cooperation. In June 1995, seven tonnes of Ukrainian zirconium tubing, a material with civil as well as military applications, was seized in New York and Cyprus. Unless border controls are improved throughout the world and modern security procedures introduced into the former Soviet Union, nuclear smuggling in future will make the banditry of Shamil Basayev seem like a childish prank.

'In 1993 the Deputy Minister of Information in Kuwait announced that they had found DU shells. It was conveyed on the Middle East broadcasting agency. Then two weeks later, he denied it. Although we hadn't found any proof of DU, we knew that there was a problem that must have something to do with the Gulf War. Doctors were reporting seeing many new strange diseases – especially in children. There was also a sharp rise in defects in sheep. Leukaemia used to be the seventh most common cancer in 1989. In 1993 it had become the fourth highest cancer. In September and October 1994 we looked again for DU. This time we looked in areas where there had been Republican Guard battles. This was further west than the main front-line battles. We found contaminated Iraqi tanks and unexploded DU rounds. Samples from destroyed tanks and armour were taken and analysed – they were found to be positive . . . We now had proof that DU had been used during the war.' Dr Layth Al-Kassab, President of the Iraqi Society for Environmental Protection and Improvement.

In the course of Operation Desert Storm the Western allies fired 6,000 tank rounds and 940,000 airborne rounds of depleted uranium (DU). Under the present UN sanctions it is hard to assess with accuracy the extent to which DU contamination has affected southern Iraq. In a leaked UK Atomic Energy Authority report, however, it was estimated that the amount of DU deposited on Kuwait and Iraq was at least forty tonnes, or enough to place half a million lives at risk, although Greenpeace came up with a figure of 300 tonnes. A number of studies point to a simultaneous increase in childhood cancers, birth defects and male infertility in southern Iraq. In the province of Basra, Dr Muna Elhassani of the Iraqi Cancer Registry has reported 56 per cent increases in leukaemia. Dr Selma Al-Taha, a consultant geneticist at the Saddam Medical City in Baghdad, claims to have seen an alarming rise in cases of genetic disease and malformation during the same period of time. And yet the war in the Gulf was often presented by American generals and spin doctors as a 'clean', almost benign war. In fact the conflict provided an opportunity for the allies to test a variety of new and highly sophisticated weapon systems, or 'smart' bombs, as they were called, because they were so precise that only military and strategic targets were destroyed while the civilian population was left unharmed. Middle America kept abreast of the television war by dint of crude video images. Viewers marvelled at the allies' 100 per cent strike rate on Iraqi bridges, chemical plants and weapon depots. Upon further scrutiny, however, it became clear that the sales pitch accompanying the weapons had the same level of success. As a result of the conflict American arms sales rocketed and the Patriot Missile and the Stealth Bomber entered the lexicon of modern warfare.

Natural uranium contains three isotopes: uranium-234, uranium-235 and uranium-238. It is the uranium-235 isotope that is required for nuclear power and nuclear weapons, because of its high fissionability. The isotope is therefore isolated by a process of separation or enrichment. This is an extremely wasteful process: only 0.7 per cent of natural uranium contains the required uranium-235. The 'depleted' uranium discarded after this enrichment process contains a mixture of uranium-238, uranium-234 and their decay products, all of which differ in radioactivity and vary in half-lives. The vast majority of the waste, however, contains uranium-238, a slow alpha emitter with a half-life of 4.5 billion years. Once inside the body uranium-238 lodges in the lungs, bones and kidneys and can cause cancer.

The DU missile was designed more than fifteen years ago in the United States in order to make use of depleted uranium. DU has a density two and a half times that of steel and one and a half times that of lead. The heavy metal is concentrated in the tip of a projectile. It thus provides the weapon with an armour-piercing capability way beyond any conventional weapon. A DU projectile has a velocity of 1,500 metres per second, more than five times the speed of sound. It also has a range of up to 5 km. During the Gulf War it was thought that the armour plating on the Russian-made T-72 tanks used by the Iraqi army was enhanced. DU weapons were therefore flown out from Britain as part of a programme known as 'Operation Jericho'. The American A-10 pilots fired 30-millimetre DU rounds at Iraqi tanks, a practice known as 'plinking', which is American slang for shooting tin cans. One report during Operation Desert Storm recorded a DU missile penetrating a T-72 tank, exiting the other side and then destroying the tank next to it.

When a DU missile hits a target it fractures, oxidizes and burns. Tiny particles of uranium-oxide dust scatter on impact, contaminating the immediate environment and thereby increasing the risk of illness due to inhalation or ingestion. Many allied troops were not briefed about the hazards of radioactive missiles. Yet, as a result of 'friendly fire', twenty-nine US military vehicles were contaminated with depleted uranium. According to the Army Surgeon General, thirty-five soldiers were inside these vehicles at the time, and more than half of them were wounded by DU shrapnel. After the war, as the vehicles were being stripped by maintenance personnel, the duty sergeant told investigators from the US General Accounting Office that he was unaware of the contamination. Depleted uranium and its risks had not been part of his training. One soldier reported that he had remained in his stricken vehicle for several days, until the ground battle was over. He explained to the investigators that he had never been told of his exposure to the depleted uranium nor had he received a medical check-up. The troops in charge of wrecked military hardware did not check the vehicles for DU. Nor were they adequately protected against it.

In the aftermath of the Gulf War, the American manufacturer of the DU 'penetrator', Aerojet, wrote a letter to the US Nuclear Regulatory Commission urgently requesting new supplies of depleted uranium. 'We are currently working around the clock, seven days a week, to fulfil US Army requirements for penetrators,' the company claimed. According to Leonard Dietz, a former scientist at the Knolls

Atomic Power Laboratory in New York State, DU missiles have revolutionized modern land warfare. He said, 'I think that it is accurate to say that the DU penetrator is the most significant battlefield weapon developed since the machine gun. The tanks and armoured vehicles of most Third World nations are now potential scrap iron.'

Used-car salesman Tom Johansen has run the Frontier Car Corral in Pocatello, Idaho, for the last sixteen years. Since high school car and truck sales have been his passion. But he has also kept an eye open for lucrative sidelines. One of those is buying scrap metal. Idaho has a number of chemical plants and military bases which have regular sales of disused equipment. Ninety per cent of the scrap Johansen buys comes from the US Government's Idaho National Engineering Laboratory (INEL), the 2,300-square-kilometre nuclear site, 80 kilometres north of Pocatello. 'Normally you get details of the sale through the post,' he explains. 'You drive out there to go and view the stuff, which is cut up and left in piles for viewing, and you place your bid.'

The auction he attended in July 1993 was a little different. The sale paperwork stated that owing to its 'unique quantity and quality', the lot was being stored in a Pocatello warehouse. 'When we went to inspect the goods, we were ushered in past armed guards, which was kind of strange. Once inside I was amazed. We all were. It was massive. The warehouse was full of this brand-new engineering equipment made of real exotic metal, wrapped in sheets of plastic.' There were 700 tonnes of it; 165 shiny slab tanks, wash columns, heat exchangers and evaporator vessels, made of the highest grade stainless steel. Johansen put a bid in at US$153,999.99. Two days later it was announced that he had been successful. He returned to the now unguarded warehouse to look again at what he'd bought. 'While I was in there, the head warehouseman asked me what I was going to do with it. I told him I was going to cut it up for scrap. He said, "Boy, before you do that, I should see if you could sell it for what it was intended." I asked him what that was, and he said that it was a nuclear reprocessing unit to make highly enriched uranium.'

The Idaho National Engineering Laboratory is one of seventeen US Department of Energy sites that contribute to America's nuclear weapons stockpile. The new processing unit was earmarked for the laboratory's work in converting the Navy's spent nuclear fuel into the enriched uranium-235. This in turn is transported to Oakridge, Tennessee, to another Department of Energy site, where it is refabricated into fuel for a reactor that produces plutonium and tritium for nuclear warheads. The project was exhausted during the 1980s at the height of Reagan's policy of re-arming America. In 1992, the Bush Administration called a stop to reprocessing on American soil, except for a handful of research facilities, citing the implications to the environment and the issue of non-proliferation as the reasons. For a time the replacement unit was stored in the Pocatello warehouse under armed guard. When the Department of Energy grew weary of forking out US$40,000 per month for something that was now a white elephant, it decided to sell the unit.

Johansen wanted to find out exactly what he had bought. He contacted the designers in California. 'The head designer was sick that he'd spent nine years of his career on this and it had ended up in my hands. He said that it was hard for me to understand, but if I looked at reprocessing like a Ferrari, I had just bought the engine.' Johansen also requested purchase orders from contacts at INEL. When he matched up what he had with the purchase orders, Tom discovered that the total cost was a few dollars shy of 16 million.

With no prior knowledge of the atomic industry and with a sales patter more suited to selling a 1972 Ambassador Sedan car, Johansen and his wife Sandy launched themselves on the international nuclear market. They had one lead. Two engineers from British Nuclear Fuels had visited the warehouse a few months before and were interested in that type of reprocessing. 'We didn't know who British Nuclear Fuels were, so Sandy rang the British Embassy in Washington, who gave us the number. We wrote a letter to a Mr Langley, who requested an inventory of what we had. So we got that to him. Then we got a call from him saying that they weren't interested at this stage, but might be later.' In fact British Nuclear Fuels were concerned and immediately contacted the Ministry of Defence, who fired off a hand-written letter to the US State Department. 'I don't know if you know, but Frontier Salvage of Idaho are trying to sell a nuclear fuel reprocessing plant! BNFL aren't interested. I wondered if Saddam Hussein et al. might be,' wrote Ray Gatrell, the nuclear safeguard official in Whitehall.

In the meantime, the Johansens had approached the Japanese, who found it hard to believe that a used-car salesman could get his hands on this type of equipment. They requested the blueprint. Johansen was advised, however, that the only way to get hold of blueprints was to file a claim under the Freedom of Information Act. Doubtful that he'd get hold of such sensitive information, he went ahead anyway. A few days later he got a letter telling him to come out to the site to pick up his blueprints. 'After that, when it became known that I had got the blueprints, engineers from different divisions out there [at the INEL site] kept calling and asking me if I wanted the test coupons, the certified radiographs, the X-rays and stuff like that. So we took the lot.' The Japanese were impressed, but where, they asked, were the flow charts? Johansen got back on the phone to INEL and spoke to Lloyd McClure, manager of technology transfer for INEL contractor, Westinghouse Nuclear Co. McClure sent him the flow charts and a government directory of nuclear facilities worldwide. 'The documents were unclassified,' McClure later explained. 'We have a charter at the national labs to help local communities and local industry be more cost efficient. I guess I never really thought at the time about the threat of it falling into the hands of some Third World country or anything like that.'

The Johansens were beginning to get more and more enquiries. They hired a marketing firm who made a short video of the equipment and the paper work. The video was sent to potential clients. Nuclear officials from Russia called. A trading company in Australia, claiming to work for the Indian Government, faxed through an offer of US$8.3 million and appended its banking details. The South African Government were also interested, but cautious. They called the Nuclear Regulatory Commission (NRC) in Washington. By this stage the NRC had heard about Frontier Salvage from the State Department. The equipment, they told the South Africans, would not be leaving the country.

Betty Wright, the NRC's export/import licensing officer, called the Johansens to tell them the equipment was not exportable. 'The idea of the NRC calling us kind of panicked me,' Sandy Johansen recalls. 'So at that point we just put a hold on everything.' She wasn't the only one panicking. Behind closed doors in Washington the 'Idaho case' was getting a lot of attention. Everyone was wondering how the hell something like this could have happened. The Nuclear Regulatory Commission wrote to Hazel O'Leary, the Secretary of Energy, to get some answers. Five months later, in July 1994, the Department of Energy let the NRC know that it was all in hand. Shortly after, O'Leary was giving a press conference where she was asked a surprise question about the 'Idaho case'. With the story now out, O'Leary assured the reporters that her department would get hold of the equipment, whether that meant buying it or seizing it. The following day security guards were back at the warehouse and two government negotiators were flown in to Pocatello. They offered Johansen US$600,000 with the option to keep the equipment as scrap. He wanted to think about it for a few days, but was told that that wasn't possible. 'That made me angry. I'd been trying to get this problem sorted for months and all of a sudden, in one day, these guys wanted to have a Chinese fire drill.' It was urgent because the press were camped outside Johansen's office. The negotiators pleaded with him to tell the reporters that a tentative agreement had been made. This he did. But a firm agreement never materialized. Johansen insisted on a few days to put it to legal counsel, and the Department of Energy withdrew the offer. Their next offer was for the amount that Johansen had bought it for originally. Johansen counter-offered with US$1.2 million, expecting the negotiators to return with a sensible figure. They never did. 'That government man was like a pouting little kid, you know. It was like I'd pulled him away from his golf game to come and make this deal.' But the Department of Energy hadn't finished. They were hoping that Johansen would get fed up waiting for an offer and would start cutting up the equipment. Johansen, however, had other ideas. Now quite an expert in reprocessing equipment, he was busy moving the essential parts to a secret location in case the Government should decide to seize it.

In November 1994 the Department of Energy flew in the undersecretary's right-hand man, Tom Todd. The first thing that he agreed to do was to take a look inside the warehouse, something that the other negotiators had never done. 'He couldn't believe it. He was just totally amazed,' Johansen explained. 'He'd seen it on TV but he had no idea of how much there was.' Negotiations got under way starting at US$300,000. By that stage Johansen estimated that he'd spent US$286,000 in total, so counter-offered with US$600,000. A week before Christmas they eventually agreed a figure. The Department of Energy would buy it back for US$475,000 and Johansen could keep it as scrap. At the beginning of January Johansen, the used-car salesman, now a local hero, started to cut it up.

Looking back, Johansen reckons it has been worth the effort. 'I probably know more about nuclear processing now than most people at INEL,' he laughs. At the last meeting Tom Todd gave him a copy of the Department of Energy's new manual of tightened regulations on how future sales of nuclear equipment should be handled. Scrap sales at INEL are postponed for the moment while employees learn these new restrictive practices. Johansen swears he'll be first in line when the sales start up again.

Epilogue

In eastern Germany, only a few kilometres from the border with the Czech Republic and within sight of the Erzgebirge mountains, lies the Zwickauer Mulde valley. The valley has a mining tradition that stretches back to the fifteenth century when it was discovered that the mountains were rich in silver. For generations the area prospered by dint of cobalt, bismuth and nickel. But over the years the valley was reduced by shallow tucks and folds until huge piles of waste began to crowd the towns of Schneeberg, Aue and Schlema. Now vast slag heaps tower over the miners' terraced cottages stopping only yards away from their gardens. The Hummerberg is a waste pile so big – 34 hectares of loose waste rock, 3.5 km long – it appears on local maps under its own name. It dominates the landscape – and yet there is more to the Zwickauer Mulde valley than meets the eye. The underground seams rich in silver and cobalt are even richer in uranium. But for centuries the radioactive ore was discarded as waste and piled high wherever space was available. It wasn't until the reunification of Germany that the full horror of this environmental disaster was recognized. Twelve hundred square kilometres are now thought to be contaminated. Unwittingly, the people of Schneeberg, Aue and Schlema have been living in this radioactive wasteland for hundreds of years. Their valley is now referred to as 'death valley' by the tabloid press, because of the high numbers of cancers and cases of silicosis. Natural background radiation is higher here than anywhere else in the Federal Republic. Radon gas (a radioactive decay product of uranium) collects in basements and impregnates building materials. Five years ago the German Government allocated 30 billion Deutschmarks (£12 billion) to the environmental clean-up of the area. Affected buildings are currently being fitted with sophisticated ventilation systems. The smaller mines are being backfilled and the Hummerberg, as with all the large slag heaps, is being covered in a thick layer of soil and planted with saplings. Understandably, the local people are in a state of shock and still reeling. Eight million cubic metres of waste have been removed in the clean-up so far, but it is estimated that at the current rate there will be little visible change to the landscape before the year 2006.

At the end of the Second World War the Zwickauer Mulde valley, having become a part of the Soviet bloc, suddenly became a classified secret. People living inside the restricted area were ordered to have a triangle stamped on their identification papers. The uranium ore discarded by the mining process was crucial for Stalin's atom bomb project. In 1947, with help from the Soviet army and the KGB, the first extraction of the ore began. The effort to break America's nuclear monopoly became a national obsession: 140,000 miners were recruited to work three shifts, twenty-four hours a day in over 400 mines. The area was the Soviets' main supplier of uranium throughout the Cold War and the third largest producer in the world. The final shipment left for Russia at the end of 1990.

I met a group of uranium miners in a *Bierkeller* in Schlema. The men, who had worked in the mines all their lives, referred to the 1940s and 1950s as the Klondike period. Many of their stories were familiar. Safety measures were ignored, they said. Inside the mines ventilation was poor and protective equipment against radioactivity was more or less non-existent. Until the mid-1960s dry drilling and explosive techniques had meant that each miner worked up to twelve hours in terrible conditions, breathing the contaminated dust. 'I was fifteen when I became aware that you could get sick working in the mines,' recalled Seidel Klacy, whose brother had just died of lung cancer. 'But nevertheless I did my twenty-five years.'

'Others were dead by the age of thirty,' I was told by Rudolph Karl-Heinz. 'But we've been in the mines too long to worry about our health now. We don't know what is inside us so we just hope that we are not the ones to die.' Most of the people I spoke to came from families with a time-honoured mining background. They had started work at the age of eighteen – and why not? Unemployment was high and it was a good career. Miners were respected in the community and the wages were almost double those at the cutlery factory in Aue.

'We got used to a higher pay and after a while you cannot suddenly accept lower wages. We simply wanted to survive and feed our families,' said Klacy. There are currently 15,000 cases of silicosis in Schneeberg, Aue and Schlema. Moreover, 10,000 patients suffer from related ailments. In 1990, 7,000 people were reported to have died from lung cancer.

According to Wittke Gunter, a miner originally from Poland, it was obvious within a year of Mikhail Gorbachev's promotion to the Kremlin that mass employment in the valley was coming to an end. But the redundancies didn't begin until 1990. Five centuries of mining in Zwickauer Mulde were over. Now 28,000 miners are out of work and the valley faces an uncertain future. The clean-up employs only 10 per cent of the workforce and, in any case, it is estimated to be finished by 2010. One idea is to regenerate 'death valley' by promoting it as an alternative health resort for sufferers of arthritis, plumbism and similar complaints. During the first half of the century it was widely believed that the radioactive waters had curative properties. In 1918 the Ober Schlema, one of the largest health spas in Europe, opened for business offering a clinic, recreational facilities and 230 beds. (It claimed proudly to have the most radioactive waters in the world.) Seventeen thousand invalids flocked each year to the Zwickauer Mulde valley right up until 1943, and poignant reminders of the boom years are evident today. Above the doors of many of the larger houses in the valley you can still make out the names of Haus Elfrieda, Haus Martina, Haus Bautzen. The spas were closed finally in 1946.

A hundred years of evidence to the contrary has not dispelled the belief that radioactivity is a cure-all. In October 1994, for example, I visited the sleepy frontier town of Boulder in Montana. The town boasts no more than a main street, a handful of pick-up trucks and a cluster of old uranium mines, yet it receives almost 10,000 visitors a year. 'They come from all over,' I was told by Pat Alverson, owner of the Sunshine Radon Health Mine. 'We get Eskimos, Indians, even pensioners from Florida.' And here is a paradox: Middle America is currently gripped by a fear of radon – the colourless, odourless, radioactive gas which many experts claim to be second only to smoking as a cause of lung cancer. Yet legions of the sick and infirm go to Boulder each year and pay money to inhale the gas. 'I've seen people hobble in here on crutches, and after a week in the mines forget to take them home with them,' says Alverson. 'To be honest I don't know how it works, but it does.' Each visitor completes an annual course of between forty and sixty hours in the mine shafts. 'My husband and I stayed ten days,' a woman from Langerly in Canada told me. 'We both feel better. My husband says that his hair doesn't fall out any more, his movement is much better, and his breathing too has improved.' In other words, the mines that once provided uranium ore for America's nuclear arsenal now compete with one another in the free market of alternative health care. The Merry Widow mine has a gift shop selling embroidered baseball caps, coffee and doughnuts. The miners' old bathhouse at the entrance to Free Enterprise mine is now a rest room with a television, a visitors' book and half a dozen beds. Deep underground, the 360-metre mine shaft has been furnished with tables, couches, exercise bicycles, books and board games. There are radioactive water pools in which to bathe hands and feet, and soft lighting to make the patients feel as comfortable as possible. The temperature is a constant 55°C.

In Sellafield, as in Hanford and the Zwickauer Mulde valley, a radioactive century is coming to an end. Yet the atomic future is hard to predict, especially in Britain where the nuclear power industry faces extinction in the wake of a decision by British Energy in December 1995 to abandon plans to build any more reactors. The U-turn sounds the death-knell for nuclear power in the United Kingdom, and its timing is instructive, not least because the future of Sellafield depends upon the THORP reprocessing plant, whose underlying principle now seems misguided, outmoded and utterly ineffective in terms of cost. The Calder Hall reactor at Sellafield was the world's first nuclear power station when it opened in 1956. It had been hastily finished in order to start up before an American rival at Shippingport, and the Queen officially flung the switch as the first nuclear-generated electricity was connected to the national grid. Electricity, as it happened, was a by-product: Calder Hall's primary function was to manufacture plutonium for the atomic bomb. As a matter of fact the United

Kingdom's first civil nuclear power station – at Berkeley in Gloucestershire – did not open until July 1962. The 167-megawatt Magnox plant closed in 1989 and is currently being decontaminated. The world we live in bears little resemblance to the one envisaged by the founding fathers of Calder Hall. There are no Panatomic Canals, no fruit gardens in the Sahara, no limited nuclear wars. In short, the Cold War is over. The heyday of Dr Strangelove is past. In future the talk will be of decommissioning nuclear weapons and reactors. Yet mankind is still wrestling with the legacy of radioactive waste, still trying to come to terms with the dark side of nuclear physics, our hubristic interference in the fundamental order of things. Perhaps an understanding of radioactivity and its half-lives, which unavoidably link the future to the past, can help resolve the issue in a way that lessens the risk of contamination. Such, at any rate, is the hope of dozens of geologists, hydrologists, physicists and chemists who gave evidence in 1995 to a public inquiry at Cleator Moor, near Sellafield, where Nirex is looking to begin preliminary work on a nuclear waste dump. If the environmental scientists are right in even the broad contours of their outlook, then there is good news and bad news for the rest of us. The good news is that forecasts of a clean-up were essentially on target. The bad news is that a decontaminated planet is not merely a hopeless ideal; it is a logical mirage.

Table of Half-lives

Element	*Isotope*	*Half-life*
Beryllium	8	1 × 10−16 second
Polonium	213	0.000004 second
Fluorine	17	66 seconds
Technetium	104	18 minutes
Potassium	42	12.5 hours
Sodium	24	15 hours
Tritium	90	64 hours
Radon	222	3.82 days
Iodine	131	8 days
Phosphorus	32	14.5 days
Calcium	45	164 days
Cobalt	60	5.3 years
Strontium	90	28.1 years
Caesium	137	30 years
Radium	226	1,590 years
Carbon	14	5,730 years
Americium	243	7,950 years
Plutonium	239	24,400 years
Thorium	230	80,000 years
Uranium	235	710 million years
Uranium	228	4.5 billion years
Rubidium	87	47 billion years

Acknowledgements

There are numerous people who have helped me write this book and I apologize to any that I have inadvertently failed to mention.

I would like to thank John Kelly, whose constant supply of information made my job a lot simpler. I also wish to thank Bill and Gloria Addington, Janine Allis-Smith of CORE, Ian Atkinson, Keith Baverstock, Dr Beschorner at the Federal Ministry of Economics, Black Hill Productions, Brennelementlager Gorleben, British Nuclear Fuels Ltd, Richard Cade, David Polden and Paul Hawkes at CND, Steve Comley of We the People, Wolfgang Ehnke, Sam Fowler, Dr Patrick Green and Dr Rachel Western of Friends of the Earth, Antony Froggatt of Greenpeace UK, the Gundersen family, the Government Accountability Project, Pete Gullerud, Caz Hildebrand, Gráinne Kelly, Sergei Kozyrev, Hugh Livingstone of the Edge Gallery, Igor Morozov, Sarah Murray, David Kyd at the IAEA, Mary Olson and Diane D'Arrigo of the Nuclear Information and Resource Service, Arjun Makhijani, Julia McGowan, the McKay family, Sherry Meddick, Jeffrey Moag of the National Security News Agency, Leroy Moore of the Rocky Mountain Peace Center, David Perry of the NRPB, US Department of Energy, Stan Reid of BNFL Inc., Betty Richards, John Rodgers, Yvonne Ryan, the Snake River Alliance, Bob Schaeffer of the Military Production Network, Elke Schlamp, Ralph Schmitt of Bundesamt für Strahlenschutz, Alexander Sich, Don Hancock of South-west Research and Information Centre, Thomas Tebb, Mike Townley, Kathleen Trauth of the Sandia National Laboratories, Kitty Tucker, Jonathan Turley, Howard Vasquez and Jay Lees of WIPP, the Westinghouse Hanford Company, Dr Werner Runge, Johannes Bottcher and Hans Simon of Wismut.

Lastly, I would like to thank my wife Sumaya for her encouragement and suggestions, support (and patience) throughout.

PICTURE ACKNOWLEDGEMENTS

We gratefully acknowledge permission to the following to reproduce photographs in this book.

Henri Becquerel © Mary Evans Picture Library
Pripyet © Igor Kostin/Imago/Sygma
Wendell Chino © Dianne deLeon-Stallings/*Ruidoso News*, New Mexico
Edward Teller © Dr Eitan Abraham
Greenpeace protest © Greenpeace/Kumagai
Nuclear train © London Region CND/Paul Aston
Map of routes © London Region CND
Tom Johansen © Eduardo Citrinblum
Mutant scorpionfly © Cornelia Hesse-Honegger
Sellafield © Environmental Picture Library/Steve Morgan

All other photographs reproduced by courtesy of the author.

Index